T0211775

A Critique of the Moral Defense of Vegetarianism

A Critique of the Moral Defense of Vegetarianism

Andrew F. Smith

First published 2016 by
PALGRAVE MACMILLAN

The authors have asserted their rights to be identified as the authors of this work in accordance with the Copyright, Designs and Patents Act 1988.

Palgrave Macmillan in the UK is an imprint of Macmillan Publishers Limited, registered in England, company number 785998, of Houndmills, Basingstoke, Hampshire, RG21 6XS.

Palgrave Macmillan in the US is a division of Nature America, Inc., One New York Plaza, Suite 4500, New York, NY 10004-1562.

Palgrave Macmillan is the global academic imprint of the above companies and has companies and representatives throughout the world.

ISBN 978-1-349-71708-8
E-PDF ISBN: 978-1-137-55489-5
DOI: 10.1057/9781137554895

Distribution in the UK, Europe and the rest of the world is by Palgrave Macmillan®, a division of Macmillan Publishers Limited, registered in England, company number 785998, of Houndmills, Basingstoke, Hampshire RG21 6XS.

Library of Congress Cataloging-in-Publication Data

Smith, Andrew F., 1972–
    A critique of the moral defense of vegetarianism / Andrew F. Smith.
      pages cm
    Includes bibliographical references and index.
      ISBN 978-1-137-55488-8—ISBN 1-137-55488-6 1. Vegetarianism—Moral and ethical aspects. I. Title.

TX392.S58 2016
179'.3—dc23   2015027282

A catalogue record for the book is available from the British Library.

For Isabel Hawman, the Schuylkill River

CONTENTS

PREFACE

For the record, I did not expect to end up as an iconoclastic vegetarian when I began work on this project. I had no intention of even writing a book when I first set out, let alone a book in which I would challenge some of my own most entrenched beliefs. The development of *A Critique of the Moral Defense of Vegetarianism* has proceeded so quickly that it will take some time for me to fully digest (pun intended) what I have written here.

A Critique of the Moral Defense of Vegetarianism comprises two stories, one of exploration and the other of discovery. Initially, I set out to answer a question that occurred to me as I read through the comments of a blog post on the website of a divisive figure in my field. The author of the post suggested that omnivores, not vegetarians, bear the burden of justifying their dietary practices, because omnivorism causes far more suffering and harm than vegetarianism does. Among the numerous responses to the post were several comments that focused on whether or not plants could suffer. They initially were offered in support of a sorry trope that vegetarians are forced to address ad nauseam. Namely, vegetarians are hypocrites because they condone killing and eating plants but proscribe killing and eating animals. In reply came a flurry of responses, including several in which it was (expectedly) emphasized that animals have the capacity to suffer—and in fact suffer greatly as a result of practices that are common today—when raised and killed for food. Plants do not, the respondents claimed. After all, in contrast to animals, plants are not sentient. This is a default assumption maintained by just about everyone who weighs in on the subject, whether in the philosophical literature or in the popular press. And since good reasons can be offered even to avoid killing animals painlessly for food, the charge of hypocrisy gets deflected.

I wondered, though, how the defense of vegetarianism would have to change if plants actually were sentient. I also questioned if I was warranted in presupposing without any real evidence that plants are not sentient in order to defend my own vegetarianism. So I dove headfirst into a body of scientific literature by researchers who study what has come to be called "plant neurobiology" and was amazed—stupefied—by what I found. When I say I was stupefied, I mean that I felt very stupid. Dumbstruck. There is a great deal of empirical evidence to support that plants are indeed sentient. This means that a common moral defense—my go-to defense—of vegetarianism fails, as I address in Chapter 2. So I went ahead and developed an alternative defense and thought I had a halfway decent journal article on my hands.

The thing is, I have trouble separating my research from my personal life. And I am getting worse at doing so as time goes on. I have developed the habit of processing my own baggage through my professional writing. So it wasn't enough that I had what I thought was a publishable essay. I had to spend much more time considering how my own practices would have to change. This is when things got complicated, when cracks in my self-understanding and sense of my place in the order of things began to appear. I realized that the new and better defense of vegetarianism that I had developed still was not good enough. I faced bigger, much bigger, problems. That my problems were—are—much bigger did not necessitate a particularly long book. Sometimes even complex issues can be unpacked fairly expeditiously. My difficulties at this point instead have to do with implementation, with actually heeding my conclusions. But I suppose this puts me in good company with quite a few past and present philosophers.

So that is the beginning of my story of exploration. I invite you to continue reading if you would like to see the plot unfold. I now provide a brief sketch of my story of discovery—a story whose development has been even more surprising to me than my story of exploration. It also is a very personal story, meant more for me than for you, to be honest, that I had no idea I needed to tell. No, that is not quite right. I was not aware that there was anything to tell until my initial article-length project began to morph into a book.

In November 2007, the day after I defended my dissertation, my maternal grandmother died. I did not visit my Grandma Hawman all that often. Even when I was a child, we lived a day-long car drive from her home in eastern Pennsylvania. But she nevertheless was a powerful figure in my life. She was the genuine matriarch of my mother's large family, and she represented something of an anchor for me. This is hard for me to explain, because it is hard for me to understand. Permit me simply to say that I felt deeply connected to her. The world at least made some sense as long as I knew I could enjoy a warm afternoon chatting with her on her front porch or spend an evening with her playing card games.

Then she was gone. I went from the high of my dissertation defense to her funeral in a matter of days. And I never properly processed her death. I am not sure I ever mourned her loss, although I did not realize this until my story of discovery associated with this book began. All I can say is that I have been at sea in a number of ways since she passed away, and I did not realize just how much her death contributed to this. I am happy now to be piecing things back together. And while my discovery is in no way a final piece of the puzzle, it most assuredly is an important one.

I now live in Philadelphia just a stone's throw from the beautiful Schuylkill River. I run almost every day in Fairmount Park along its banks. I love this place. The city has been very welcoming, for which I am so thankful. And no part of it has been more welcoming than the Schuylkill. Now I know why. You see, my Grandma Hawman is buried in a cemetery upstream in Reading that also is a stone's throw from the Schuylkill. Yes, that is right. All along, it has been my grandmother, who is now interlaced with the river, who has been so welcoming. I did not, could not, see this for years. And while our connection is incomplete for reasons that I explore later, I finally have begun to mourn her death. Because I now see that she was never gone. I hope this makes more sense as the book unfolds.

Oh, and I have one more thing to mention about my Grandma Hawman. I never told her that I was a vegetarian. I thought she would not understand, that she would find it difficult to know how to feed me when I visited. So I strategically scheduled visits to avoid meals! The thing is, I now see that I had much more to learn than she did. And I have learned a lot in writing this little book. I hope that you do, too. I look forward to hearing what you have to say about it.

Acknowledgments

Because this project came together so quickly, few people had a direct hand in its creation. But that I acknowledge only a handful of people should not be taken as an indication that I basically wrote it alone. I simply doubt that my publisher would condone listing a lifetime full of people whose support—and opposition—has been integral in the formation of the ideas that you find here.

For their inspiration, I must thank Daniel Quinn, Val Plumwood, Matthew Hall, Michael Marder, Graham Harvey, Melissa Nelson, Derrick Jensen, Aric McBay, and Deborah Bird Rose. Reading their work has been life-changing for me, literally. Daniel Quinn's direct and detailed feedback on an early draft of the entire manuscript was both most welcome and incredibly helpful. Naomi Zack provided very helpful comments on the second chapter and pointed out to me a crucial hole in my argument for the transitivity of eating. Moreover, it is she who recommended that I consider transforming the project into a longer work.

I thank an anonymous reviewer of the manuscript for extremely thought-provoking comments. Your enthusiastic support means a lot to me. I am grateful to Abioseh Porter for his mentorship and friendship and to Donna Murasko for the generous release time she has provided for this project and others. The editorial and production teams at Palgrave Macmillan have been a pleasure to work with. They are consummate professionals and genuinely nice people. I thank Phil Getz for the strong support he has provided from day one, Alexis Nelson for amicably handling my many niggling questions, and Rachel Crawford and Daniel King for smoothly taking the book through production.

I give a shout-out to my neighbors in West Philly. Thank you for helping make a new house into a home. I also thank the Schuylkill River Trail and all those who maintain it. You have become a permanent fixture in my life, and I would have it no other way.

I thank my parents, Fred and Judy Smith, for a lifetime of love and support. I have not always recognized them for the wonderful people they are. I have some work to do to correct that. My four cats—Celie, Hugo, Sasha, and Loki—mean the world to me. I dare anyone to deny that they are people!

Finally, I offer my gratitude to my amazing wife, Sherrilyn Billger. She is one of the most caring, most perceptive, and most intelligent people I have ever met. The copious comments she provided on the manuscript and the long discussions about it through which she persevered have been invaluable. Thank you, Hon.

CHAPTER 1

An Unsettling Question

In the latest iteration of my undergraduate Philosophy of the Environment course, I began by teaching *What We Leave Behind* by Derrick Jensen and Aric McBay. To my relief, no one dropped the class in response to our in-depth discussion of shit on our first day with the book. Yes, you read that correctly. Jensen and McBay spend an entire chapter on human excrement, the act of its expulsion, and our peculiar cultural hang-ups about both. They pull no punches either with respect to the detail of their descriptions or their willingness to speak openly and honestly about this strangely taboo subject. It was a rude introduction to the course, but my students fared admirably. I must give credit where credit is due.

We all struggled, though, when approximately halfway through the book, Jensen and McBay ask us what we should do "if—or when—we realize that this world would be better off had we never been born, or having been born, if we were to die" (2009, 199). We had long since learned that the authors have a flare for the dramatic, but this question struck my students as overwrought. It was just too much. They balked at answering it, aside from engaging in a sort of "It's a Wonderful Life" thought experiment that is not what the authors have in mind.

Better off Dead?

For reasons that will become clear as we proceed, the question does not strike me as overwrought. Maybe this is the case for you, too. But even if you do side with my students (I hardly blame you if you do, especially since I have given you precious little context), I hope you will indulge me as I give some thought to how best to answer it.[1]

In suggesting that the world would be better off without us, Jensen and McBay's point is not that human beings are inherently destructive. We cannot possibly be or our species would have died off long ago. But many of us, including yours truly and anyone reading this book, now inhabit a culture that is inherently destructive. Indeed, this culture of ours is nothing short of ecocidal. It operates, if implicitly, according to a set of malignant axioms identified by Daniel Quinn. Here are four that are especially useful to highlight for our purposes:

1. We—the people of this culture—are what humanity is meant to be. There is only one right way to live: our way.
2. The world is made for us. It belongs to us to exploit for whatever purposes we see fit.
3. *"Humans belong to an order that is separate from the rest of the living community.* There's us and then there's Nature. There's humans and then there's the human environment"* (Quinn 2007, 174).
4. The world is "endlessly and infinitely resilient," as Jensen puts it (2006, 299), so we can do whatever we want with it and never pay a price.

Our culture may have its flaws. So do we as individuals. But this does not impinge upon the fact that the way we live is the right way to live, the best of all possible ways to live. We are, or should be, more fulfilled than people ever have been before. Look at all the creature comforts that we have at our disposal. Look as well at what we have achieved by making good use of what the earth provides for us. We take what we want from it, and it just keeps on giving.

To be sure, notes Quinn, a good many of us have changed our views about the earth's resiliency. We have a much clearer sense of the damage we do. Moreover, people rarely use the harsh and not-so-vaguely sexist language today that our cultural forebears so freely employed when describing our attempted conquest of the earth. We no longer talk of making ourselves "the lords and masters of Nature," as René Descartes does (1985, 143), or echo Francis Bacon's calls for "putting Nature to the rack and extracting her secrets" and "storming her strongholds and castles" (Farrington 1964, 62). Indeed, I would guess that you disavow ideas like these. I do. But this does not alter one whit that we are embedded in a culture that operates as if they are true. In this respect, the statements by Descartes and Bacon are not anomalies or anachronisms. "These people merely articulated, brilliantly, urges that are woven together throughout our culture like rivulets in sand," Jensen concludes (2000, 20).[2]

But Jensen's conclusion suggests only that the world is better off without our culture, not that it is better off without you and me. Why countenance the latter proposition? One answer is that you and I face daunting structural barriers to acting contrary to our culture's axioms. Another answer is that the vast majority of us exist only because our culture is so proficient at and voracious with resource extraction. I have more to say about both claims and will do so when the time is right. For now, permit me to assume that Jensen and McBay are correct. If only for the sake of argument, let us presuppose that the world is indeed better off without us. This being so, we have at least three options for how to respond. We can commit suicide, ignore our presumptive anxieties about the meretriciousness of our lives and go on living as we have, or actively take steps to transform our lives so that our continued existence benefits—or at least does not materially harm—our world. I will not bother lying to you. I have given serious consideration to option one on several occasions, and I spend far more time than I like to admit immersed in option two. But on my better days, I think I do have the courage to pursue option three. I hope I do, anyway, even if I fail miserably—which I have no trouble admitting is a distinct possibility.

Why consider, and reject, option one? Suicide is a taboo subject in our culture, too, of course, which undoubtedly affects the way I think about it. But as odd as it may be, one of the main reasons that I have not committed suicide is intimately tied to shit. The people of our culture typically deposit our excrement in the worst possible place, water, rather than where it can do the most good, the land. When properly deposited, human excrement is "a gift from us to our habitat," proclaim Jensen and McBay, "a gift of fertile soil, given in response to the nourishment our habitat gives us" (2009, 5). The same can be said about our bodies when we are laid to rest upon death.

Or at least this was the case until fairly recently. Today, our excrement routinely contains unmetabolized pharmaceuticals that kill microbes living in the soil. Our bodies are laced with hundreds of hazardous chemicals, heavy metals, and other pollutants. So we cannot even give the gift of our deaths to the world without passing on a lifetime of exposure to low-level toxins, including paint thinner, dry-cleaning chemicals, wood preservatives, toilet deodorizers, gasoline byproducts, cosmetic additives, rocket fuel (yes, *rocket fuel*), termite poisons, fungicides, mercury, and flame retardants. "Today, we all live in the same chemical ocean with better and worse eddies," Florence Williams declares (2012, 208). Every gift we have to give comes at a steep price. We are junk food for the soil and its manifold inhabitants. I can think of no better way to put it.

Dying a few decades from now rather than today will not change this. Our fate, yours and mine, is sealed in this regard. So we instead can take option two and make the most of the seemingly endlessly available (and also toxic) creature comforts provided by our eminently destructive culture. For those with money, there are always "toys and more toys and more toys," state Jensen and McBay (2009, 200). But why not at least give a little thought to what option three might entail? Let us consider what it would be like to be of benefit to our world—to the land and fellow members of the community of life of which we are inextricably part. It is possible for us both to become less toxic and, perhaps more importantly, to take steps to save others from our fate.

What exactly does a critique of the moral defense of vegetarianism have to do with this undertaking? This critique has quite a bit to do with it, actually, but perhaps not exactly in the way that you expect. It is a commonplace assumption among vegetarians, and some omnivores, too, that eating a plant-based diet is healthier for both people and the planet. Even a cursory assessment of factory farming, or what Carol Adams aptly calls the "animal-industrial complex" (1993, 200), lends considerable credence to this proposition. Nowhere is this more obvious than in the United States. Factory farming here is a pharmaceutical and ecological abomination—and an international aberration. The industry in the United States is rightly designated a pariah. Nearly every other industrialized nation refuses to import American beef, for example (Fairlie 2010, 9).

But once we set aside the obvious problems associated with factory farming, the case for vegetarianism is less cut and dry. This certainly has surprised me, and it was not the only surprise that I received as I gathered material for this book. I have been a vegetarian for nearly two decades. So imagine my shock when I was forced to conclude that vegetarianism is morally indefensible. Now imagine my utter disbelief when it first occurred to me that—from a perspective that helps us to see more clearly how we can be of benefit to our world—it is actually *impossible* to be a vegetarian! Let me be blunt. I am not saying that being an omnivore is better for the planet. I am saying that *we have good reason to reject that one can be either a vegetarian or an omnivore*. It is not that the distinction itself is harmful for the planet. Rather, the distinction is emblematic of a way of thinking about ourselves and our place in the world that is deeply rooted in our ecocidal culture. To continue to think and act is if the distinction between vegetarianism and omnivorism matters implicates us in perpetuating this culture. I did not realize this when I first began this project. As I said in the preface, I find myself uncomfortable with being an iconoclastic vegetarian. But I suppose this is an inapt label if I cannot be a vegetarian in the first place.

We should not get ahead of ourselves, though. We have considerable ground to cover before we reach the point at which I can even try to convince you of this conclusion. In the meantime, I must evaluate the strongest possible cases for vegetarianism. It would not do to flail away at a straw man or even what Robert Talisse calls a "weak man": a distorted presentation of a moral or political position that one opposes. A "weak man" is a distortion because it relies for its purported success on convincing others that the opposed view is unreasonable, nonsensical, or downright crazy (2009, 169). But embracing vegetarianism is not unreasonable, nonsensical, or crazy—even if it is morally indefensible and ontologically illusory. Evaluation of these cases must be carried out with a clearer sense of what living a life that is worthwhile for the world entails. By saying this, I do not mean to insinuate that there is one right way to live. Like Quinn, I regard this proposition as malignant. But, as he also points out, there are ways to live that work for the planet and ways to live that do not, and what works for the planet also works for people in general and on the whole.

Schematic of the Argument

I also noted in the preface that I began my work on this project by seeking to establish a moral defense of vegetarianism that would stand up to the best available evidence and arguments. I hope that this intention is borne out by the two respective defenses on offer in Chapters 2 and 3, even if they both ultimately fail. In Chapter 2, I challenge perhaps the most common defense of vegetarianism, which I label the "sentientist argument." Sentientists focus on a key characteristic that (both human and other-than-human) animals exhibit—and plants purportedly do not—that should lead us to regard as morally impermissible killing and eating them. Namely, animals are sentient. Sentience is sufficient for moral standing. So animals are deserving of moral standing—indeed, *equal* moral standing— with respect to human beings. We do not countenance killing and eating human beings, and we should not countenance killing and eating animals either. No sentient being should be subject to this fate.

If sentientists give any thought to whether plants are sentient, they rule out this proposition posthaste. If they grant plants moral standing for other reasons, they generally mitigate the consequences of doing so in order to ensure that any potential conflicts of interest between us and them end up weighing in our favor. I began my work on Chapter 2 by questioning whether these considerations about plants are justifiable. I might have once felt comfortable following the general trend and simply assuming that plants are not sentient, but no longer. That which gives its life so that I have the energy I need to live deserves no less.

In defiance of both the philosophical literature and the popular press, I found more than enough scientific evidence to conclude that plants are sentient. A basic premise of the sentientist argument is thus false, which means that the argument itself collapses. So if an attempt to salvage a moral defense of vegetarianism is to be made, it must proceed on different terms. In accordance with what I call "expansionary sentientism," I argue that plants are deserving of equal moral standing because they are sentient, too. Yet we nevertheless have reason to subsist on plants rather than on animals, so long as plants are treated with due respect in the process of cultivating, harvesting, and eating them. By adhering to these dictates, expansionary sentientists do a better job of delivering on the basic principles that motivate sentientism than do sentientists themselves. Most notably, adherents of the former do considerably less harm to sentient beings than the latter do. Expansionary sentientists also acknowledge and consider how best to negotiate colliding agendas that are bound to arise between eaters and eaten.

I soon realized, though, that expansionary sentientism has irredeemable problems as well. It retains vestiges of sentientism that threaten to undermine the equal moral standing not just of plants but also of animals. Like sentientists, expansionary sentientists retain an unstated human-centered view. This leads expansionary sentientists to assume that beings *like us* are most worthy of being granted moral standing. Expansionary sentientists also assume without any further thought that we humans have the right—indeed, the duty—to confer moral standing on deserving other-than-human beings. I challenge both propositions in Chapter 3.

With these issues in mind, I argue that a human-centered defense of vegetarianism should give way to a *land-centered*, or topologically contextual, defense. The latter works far better for the planet as well as for people—and not just for human people. Drawing on considerations offered by Val Plumwood, Graham Harvey, and others, I highlight the moral significance of our animality and ecological embeddedness. This leads me to abandon the sentientist conceptual framework in favor of an animistic conceptual framework. Animists maintain that the world is full of people, only some of whom are human. Plants and other-than-human animals may be people, too. Moreover, we maintain a special bond with, and specifiable obligations to see to the needs and interests of, our landbase and all the people who live on and in it. Accordingly, vegetarianism is morally defensible so long as it works for our landbase and for fellow inhabitants of it—so long, that is, as it is ecologically and socially sustainable.

In Chapter 4, I nevertheless am forced to conclude that a contextualist approach to the defense of vegetarianism ultimately does not qualify as a moral defense of vegetarianism. More importantly, I lay out my case against the possibility of being a vegetarian, full stop. Every member of the community of life, including each one of us, is both an eater and—eventually—eaten. Actually, you and I are being eaten as we speak, as microbes consume our dead skin cells. Moreover, we are all part of the same nutrient cycle. Animals obviously eat plants, but are you aware that plants also eat animals? Rich, black soil is composed in large part by the decomposed bodies of both animals and plants. With the help of microbes in the soil, plants feed on the constitutive parts of these bodies.

So far so good, right? Let me add this: As the well-worn saying, or my variation of it, goes, we are who we eat. And our food is who our food eats. By virtue of transitivity, we must be who our food eats as well. Because plants partially subsist on animals, even those of us who assume that we subsist entirely on plant life also eat animals. This is why it is impossible to be a vegetarian. And because we are all part of the same nutrient cycle, all part of the same ongoing process of energy transfer, it also makes little sense to suggest that we are omnivores.

In Chapters 5 and 6, I take the opportunity to respond to several anticipated objections and complaints. Chapter 5 addresses the objections. There are two. First, perhaps I have overlooked or neglected circumstances that favor an ecological defense of vegetarianism, which perhaps can be further cashed out in moral terms as well. I do not rule out that this may be so. But I draw on Simon Fairlie's detailed comparison of livestock permaculture with vegan permaculture in the United Kingdom to show that matters are considerably more complicated than proponents of this objection realize. Indeed, at the end of the day, the ecological distinguishability between these respective permacultures may not be all that pronounced. Second, Mark Boyle argues that questions regarding the ecological viability of our alimentary proclivities must be circumscribed by a wider discussion of problems associated with human overpopulation and overconsumption. In assessing who and how we eat, we must consider that our landbase almost certainly needs far fewer of us consuming much less of it. Boyle is right, and his objection indicates that human overpopulation and overconsumption likewise constitute serious ecological problems that must be addressed via ecologically focused solutions. Somewhat tangentially, I bring Chapter 5 to a close by acknowledging that we may have reason to honor certain aspects of what motivates people to be vegetarians even if we can neither defend vegetarianism on moral grounds nor actually be vegetarians.

My response to the anticipated complaints—three of them—comes in Chapter 6. They reflect how I envision my students reacting upon being presented with my overarching conclusions and the argument on their behalf. Perhaps you will voice them, too. Only time will tell. First, I risk alienating vegetarians, who otherwise represent potential—if sometimes uneasy—allies for me. Specifically, I fail to appreciate that the embrace of vegetarianism fosters nonviolence and peaceful coexistence between humans and other-than-human beings. Second, despite the fact that I call for a wholesale change in how we view our relationship with who we make our food, I provide ammunition for omnivores to defend refusing to alter their dietary habits. And third, I should be satisfied with the widespread abandonment of factory farming, since this is a position that appeals to most vegetarians and many omnivores. Suffice it to say that I have something to say in response to each of these complaints, but I leave these matters for the end of the book. Let us now begin to familiarize ourselves with the fascinating lives of plants.

CHAPTER 2

PLANT SENTIENCE

Philosophers who defend vegetarianism on moral grounds tend either not to consider whether plants deserve moral standing or to reject the proposition outright. Like all living organisms, we must eat *something* to survive. By definition, vegetarians do not endorse conscientious suicide by willful starvation. But why morally countenance eating plants?

The most common answer to this question is supported by philosophers who endorse a wide array of moral theories, including consequentialism (Singer 2002 and Gruzalski 2004), deontology (Taylor 1986, Regan 1997 and 2004, Francione 2000, Pluhar 2004, Deckers 2009, and Curnutt 2011), virtue ethics (Shafer-Landau 1994 and Hursthouse 2011), contractarianism (Bernstein 1997), care ethics (Curtin 2004), "cosmic holism" (Steiner 2008), and other less doctrinaire outlooks (Sapontzis 1987, DeGrazia 1996, Engel 2000, and Fox 2006). It boils down to this:

1. Sentience is a sufficient condition for having moral standing.
2. Having moral standing entails that one should not be killed and eaten by moral agents.
3. Many animals are sentient, so they have moral standing. It is morally impermissible for moral agents to kill and eat sentient animals, and there are good reasons to reject eating *any* animals whatsoever.

4. Plants are not sentient, so they do not have moral standing in the relevant sense. It is morally permissible for moral agents to kill and eat them.

Let us call this the *sentientist argument.*

Arguably the most salient characteristic of sentience is the capacity for subjective awareness of sensations and emotional states that are pleasant or unpleasant (DeGrazia 1996, 99). Specifically, sentient beings have interests, preferences, and cares associated with avoiding pain, fear, and anxiety (Rowlands 2002, 11). They have unpleasant sensory experiences associated with actual or potential tissue damage, emotional responses to perceived threats to their physical or psychological well-being, and the desire to evade both. According to sentientists, plants are not subjects of these experiences. This is why they do not have moral standing in the relevant sense. Sentientists may have other reasons to value plants or even count them as deserving of moral consideration—because they are part of a flourishing ecosystem, say. But if what counts to determine whether we can kill and eat them is strictly their sentience or lack thereof, any other reason to give them preferential treatment is dietarily beside the point.

Yet what if plants are sentient? What if sentientists mistake our deafness to this fact for plants' dumbness? In this chapter, I provide ample evidence to support the proposition that this is indeed the case. As a result, I contend that we should reject the sentientist argument. It fails to give us justification to kill and eat plants, so it provides an untenable moral defense of vegetarianism.

The view that plants are sentient is hardly new. It is a common view among animists that dates back many millennia, it is a central principle of Jainism, and it is defended by Aristotle's student and subsequent director of the Lyceum, Theophrastus (Hall 2011). But it has been out of favor even among plant scientists until quite recently. Indeed, I rely on the work of plant scientists in this chapter because of the conceptual weight sentientists attribute to science. I base my case in favor of plant sentience on recent findings specifically by so-called plant neurobiologists, who investigate how plants experience phenomena within their environment and convert these experiences into electrochemical signals that permit rapid physiological, morphological, and phenotypic adjustments (Barlow 2008). Plants do not have neurons, synapses, central nervous systems, or brains. But they do possess a homologous information-processing system that integrates incoming data—on light, water, gravity, temperature, soil structure, nutrients, microbes, herbivores, and other plants—and coordinates behavioral responses. And plant neurobiologists offer ample reason to accept that these are not simply automatic reactions to specifiable stimuli.

Plant neurobiology is the subject of considerable controversy among plant scientists (see Alpi et al. 2007 and Firn 2004), but it is not a fringe view.[1] Moreover, the manner in which sentientists conceptualize moral standing indicates that they should give plant neurobiologists the benefit of the doubt. To do so does not in and of itself entail giving up on the moral defense of vegetarianism. On the basis of what I call an *expansionary sentientist argument*, for lack of a better label, we can attribute equal moral standing to plants and animals and still embrace vegetarianism. We have good reason not to kill and eat animals and can countenance killing and eating plants. But how plants are cultivated and harvested for consumption must change if we are to treat them with the respect they are due as sentient beings.

SENTIENTISM AND THE MORAL STANDING OF PLANTS

Sentientists have done a wonderful job of providing evidence— indeed, overwhelming evidence—in favor of the sentience of many other-than-human animals. This proposition has clear behavioral, physiological, and evolutionary support, so I will not rehash it here (Singer 2002, Regan 2004, and Rachels 2011, 884). But permit me to highlight a few key points associated with the capacity of animals to be subjectively aware of pain. Having a central nervous system is generally taken to be necessary to experience pain (DeGrazia 1996, 111). Animals who feel pain tend to avoid or escape noxious stimuli; to limit the use of injured body parts to permit healing; and, among social animals, to seek assistance from others. All vertebrates and many invertebrates produce endogenous opioids, or self-generated analgesics. Perhaps these have a function other than alleviating pain. "But this seems uncomfortably like grasping at straws," Mark Rowlands states. "By far the most natural explanation of the presence of endorphins and other opiates in nonhuman animals is that these animals can feel pain, and the endorphins are there to help control it when necessary" (2002, 7). This is a key denotation of sentience, and no sentientist disputes Rowlands's conjecture.

Most other-than-human animals seem to lack the full range of cognitive and imaginative powers humans enjoy, including the capacities to worry about the future and contemplate their mortality. Having these capacities may result in greater suffering by humans, but this is far from clear. Bart Gruzalski notes that these capacities instead may lead humans to fail to appreciate what is present to our senses in the way that many other-than-human animals do. Animals may well feel confinement, transportation, slaughter, and even routine veterinary checkups far more keenly than we tend to consider, since "there are no future-oriented distractions to mitigate the corresponding pain and fear" (2004, 130). So even if other-than-human animals do not suffer exactly like humans, they must not be denied moral standing. Sentientists readily acknowledge this.

Interestingly, sentientist icons Peter Singer and Tom Regan both suggest that killing other-than-human animals painlessly—via humane slaughter techniques, for example—also should be prohibited. For Singer, this invites using other-than-human animals merely as means to fulfill trivial gastronomic desires and becoming desensitized to their premature deaths (2002, 159f.). While Regan argues in earlier work that it is subjects-of-a-life who have inherent value rather than sentient beings per se, he more recently expresses support for granting this status to the latter (1997, 110, and 2004, xvi).[2] Moral agents have a categorical duty not to harm inherently valuable beings by means of either infliction or deprivation. Taking the lives of inherently valuable beings or denying them material conditions for life both constitute harm. This is so whether or not they experience pain in the process (1983, 117).

Just as the sentientist argument can be generalized despite the multitude of ways in which it is presented and defended, a generalizable conception of moral standing can be offered as well. It relies on at least three key principles:

1. *Principle of nonmaleficence.* Moral agents have a prima facie duty not to harm moral patients. This includes a prohibition against intentionally or unintentionally causing suffering or being party to exploitation, cruelty, or callousness (Regan 1983, Singer 2002, and Llorente 2009). In cases in which doing harm is unavoidable, such as when its commitment is necessary for one to avoid starvation, moral agents must make every effort to limit the harm they do as far as is logistically possible (Matheny 2003, 505).
2. *Principle of equal consideration.* Moral agents have a prima facie duty to treat relevantly equal interests of moral patients equally (Hoff 1980, Rachels 1983 and 2004, VanDeVeer 1983, Sumner 1988, DeGrazia 1996, and Singer 2004). This may be because moral patients are assumed to have equal inherent value (Regan 1983 and Francione 2000) or because all moral patients have the equal capacity to suffer (Rachels 2004, 877). Some sentientists qualify their application of this principle in ways that give preference to humans in cases of conflicts of interest with other-than-human animals, but I leave this matter aside (Singer 1979, 152, and Regan 2004, xxix).
3. *Precautionary principle.* Lacking full information about whether other beings count as moral patients does not provide a legitimate basis for performing actions that potentially harm them. The burden of proof for so acting falls on moral agents, who must err on the side of caution (Regan 1983, 416f., and Singer 2002, 174).[3] As Evelyn Pluhar remarks with respect to marginal cases, "Considering how high the stakes are for such beings, i.e., suffering and death, it seems that we are justified in extending to them the benefit of the doubt" (1992, 192).

Sentientists do not deny that moral agents may have compelling reasons to care about the well-being of plants. But they reject either that plants have moral standing or, if they do, that it is equal to that of animals. None of the three principles applies to how plants are to be treated, since they are not regarded as sentient. So zoocentrism is a justifiable moral stance.

Singer (2002, 35) contends that there is no scientifically credible evidence suggesting that plants feel pain (see also Fox 1999, 112 and 156). They have nothing resembling a central nervous system, and we have no reason to presume that a sessile being—a being affixed to the soil—should have evolved the capacity for pain. What good would pain do if plants cannot escape it? David DeGrazia asserts that because plants have no aversive states, they have no interests. "It would be neurotic to worry about whether running through the park harmed the blades of grass underfoot" (1996, 228). Deane Curtin remarks, "It is possible to suggest that plants experience 'pain' when they are harvested, but I think this stretches any real meaning of the word 'pain' to the breaking point" (2004, 279). And Gary Francione contends that plants have no interests because they have no perceptual awareness. Without perceptual awareness, they cannot desire, want, or prefer anything (Marder and Francione 2013).

Some sentientists, including Philip Devine, grant plants qualified moral standing. Plants do have interests—in receiving adequate light and water, for example—so they can be harmed, Devine contends. "But I take it as conceded that it would be perverse to treat the need of plants as of equal importance as the needs of human beings, even when the need in question (for water, say) is of the same general kind" (1978, 490). According to Mylan Engel, it is wrong for moral agents to treat plants as means only. Nevertheless, "Other things being equal, it is worse to kill a conscious sentient animal than it is to kill a plant" (2000, 861; see also Taylor 1986, 295).

Jan Deckers maintains that killing animals is more problematic than killing plants because the former exhibit a greater dislike for being killed than the latter. Animals "either defend themselves or flee when their lives are threatened. When plants are threatened, this is less clear as plants do not seem to possess the capacity to resist being killed to the same degree that animals do. The threat of being killed does not appear to mean much to them" (2009, 588; see also Porphyry 1989, 70f.). Plants lack the ability to distinguish between harmless and potentially harmful stimuli, Deckers asserts. They respond similarly to both. This indicates that plants are less aware of their surroundings than are animals, so "it does not mean much to them to be controlled by external factors" (ibid.).[4] By comparison, animals can rapidly and flexibly adapt, which indicates not just the capacity to take in and process more information about their surroundings but also that they are clearly averse to manipulation.

EVIDENCE OF PLANT SENTIENCE

These statements may not represent the stance of every sentientist, but they do reflect a general tendency. That this is so is unsurprising. Indeed, it is expected. On the one hand, the treatment of plants throughout the history of Western philosophy largely reflects Aristotle's *scala naturae*, or the "Great Chain of Being" (Hall 2011). On the other, scientists have provided scant evidence on behalf of plant sentience until quite recently. Sufficient evidence now exists to correct past mistakes.

This is not to say that the lived experience of plants is somehow the same as that of animals. Among other differences, their lives unfold in a much slower time frame than ours. So while plant neurobiologists use terminology to characterize plant behavior that is usually reserved for animals, they do not insinuate that we should regard the two as experientially interchangeable. This is immaterial, however. So long as we have sufficient evidence that plants behave as we expect sentient beings to behave, sentientists go wrong by denying them equal moral standing.

Sensation, Pain, and Homologous Systems

Daniel Chamowitz states that while the sensuous experiences of plants and animals differ qualitatively, due in part to the fact that plants have no discernible organ of perception, plants have sensuous capacities that are at least functionally comparable to our five senses. "Sight is the ability not only to *detect* electromagnetic waves but also the ability to *respond* to these waves" (2012, 24). Plants may not translate light reflecting off objects into images, but they do convert it into cues for physiological, morphological, and phenotypic development. They not only distinguish light from shadow but also track seasons and the time of day. It is no particular advantage for plants to see like we do, but it certainly benefits them to know the direction, intensity, and duration of light sources and potential obstructions to accessing it.

The only difference between taste and smell is that the former is dependent on solubles (by dissolution or liquefaction) while the latter depends on volatiles (on the inhalation of gases). Plants sense and respond to volatiles in the air and to solubles on their bodies. So they smell and taste. They are highly sensitive to pheromones, just as we are, using them for communication and signaling.

Plants also are able to hear. Heidi Appel and Reginald Cocroft (2014) have observed plants release defensive chemicals upon playing a recording of a caterpillar biting a leaf. One means by which plants distinguish self from other—yes, they exhibit indicators of an internal sense of self—is through the recognition of unique oscillatory signals. The movements of each plant register on a frequency that is not shared by other plants (Gruntman and Novoplansky 2004).

Also like animals, the sense of touch is electrochemically mediated in plants. Plants' sense of touch is highly developed (Trewavas 2005). Roots know when they encounter solid objects, and bushes and trees alter their morphology in response to high wind exposure. This is an evolutionary adaptation that increases their ability to survive perturbations.

Plants do not need a central nervous system to have these sensuous experiences. Do they need one to experience pain and suffering? The answer to this question is critical for assessing whether plants are sentient. Chamowitz remarks, "Perhaps 'pain' for a plant could be defined in terms of 'actual or potential tissue damage,' as when a plant senses physical distress that can lead to cell damage or death" (2012, 139). They do respond physiologically to leaf punctures from insect bites, being burned, and drought conditions in ways that are consonant with self- and kin preservation. "But plants do not suffer," Chamowitz continues. "They don't have, to our current knowledge, the capacity for an 'unpleasant' [. . .] emotional experience" (ibid.). Their responses to wounding are not the equivalent of an *ouch*, even if these responses do entail the modulation of their development. Indeed, plants do not have nociceptors: neurons that specialize in sensing noxious stimuli. Nociceptors exist to enable their hosts to withdraw from dangerous situations and alert their hosts to internal physical problems (Rachels 2004, 77f.).[5] Without nociceptors, it is not clear that these experiences are possible, Chamowitz concludes.

Why then do plants excrete endogenous opioids, most notably ethylene, when wounded or subject to stress (Buhner 2002, 197)? Recall Rowlands's claim that presuming that this phenomenon has a function other than alleviating pain "seems uncomfortably like grasping at straws." Moreover, why are plants rendered unconscious by the same anesthetics that work on animals (Pollan 2013)?[6] How they respond to insect bites, fire, and drought may give us an answer. Plants cannot escape noxious stimuli, but they do send signals indicating that the environment has become dangerous. So perhaps some sort of pain analogue plays a communicative role within and between plants, warning other parts of the plant as well as neighboring plants to make physiological adaptations to increase their chances for survival. Such an analogue need not be qualitatively equivalent to pain in order to be functionally equivalent to it. This is what Chamowitz suggests at any rate (2012, 68; see also Dicke and Bruin 2001, Weiler 2003, Trewavas 2005, and Barlow 2008). So while we cannot say that an *ouch* has occurred, it is not at all clear that we need to in order to establish that plants take definitive steps to respond to actual and potential dangers. This indicates that Deckers is incorrect to suggest that they are passive to manipulation. They continually scan their environment through the use of at least as many sensuous parameters as animals, and they take immediate steps to protect themselves and others when they register threats. The experience of an *ouch* seems unnecessary, in any case, to grant plants moral standing according to sentientists' own standards. If sentientists acknowledge that other-than-human animals need not suffer exactly like humans do to be granted moral standing, the same surely must hold for plants—so long as they too qualify as sentient.

This process of scanning and reacting is facilitated by plants' electrochemical information-processing system, which is homologous to animals' neural network. When a plant is wounded or experiences stress, electrical signals that are similar to action potentials running along nerves are sent from one region of a plant to another via voltage changes across cell membranes. These signals induce a rapid change in hormone metabolism. Specifically, they elevate auxin levels (Barlow 2008). Auxins are a class of plant hormone with morphogenic qualities. They induce cell differentiation and regeneration at the site of injury and are secreted from cell to cell along sieve tubes, elongated cells of phloem tissue, which are akin to synapses (Volkov 2000, Thompson and Holbrook 2004, and Trewavas 2007).[7] This system thus has neurotransmitter-like characteristics (Dziubinska 2003, Baluška et al. 2006, Brenner et al. 2007, Fromm and Lautner 2007, and Baluška and Mancuso 2009). "At the molecular level," Eric Brenner and his colleagues hereby note, "plants have many, if not all, the components found in the animal neuronal system" (2006, 414). According to František Baluška (2010), electrical long-distance signaling and the existence of processes akin to action potentials in plant cells and tissue support the proposition that the sensuous abilities of plants are not limited at all in comparison to animals. Indeed, plants produce several proteins found in animals' neural networks, including acetylcholine esterase, glutamate receptors, GABA receptors, and endocannabinoid signaling components.

But we should not make a fetish of neurons, Stefano Mancuso remarks. Neurons "are really just excitable cells" (quoted in Pollan 2013, 92). Plants have excitable cells, too—particularly at the root apex of the transition zone.

Perhaps someday soon plant neurobiologists will determine whether or not plants experience not just a pain analogue but an actual *ouch*. What we do know, James Rachels remarks, is that "research constantly moves forward, and the tendency of research is to extend the number of animals that might be able to suffer, not decrease it" (2004, 78). Only time will tell whether plant neurobiologists can say more definitively that plants suffer in a manner akin to how animals suffer. But in accordance with the precautionary principle, what we have seen thus far should raise some doubts about the sentientist stance toward plants. Focusing next on the adaptive advantages of sessility, modularity, and self-awareness and, after that, on evidence of plant intelligence should raise more.

Sessility, Modularity, and Self-Awareness

Journalist Helen Phillips notes that plant scientist Anthony Trewavas

> thinks the only substantive difference between us animals and our distant green relatives is how mobile we are. We're used to judging intelligence by actions, he says. It's what we do and say that reveal what goes on inside our minds. So plants, silent and rooted to the spot, naturally don't appear too bright. But they do move and they do react to the world around them, he says, and they do so in intelligent ways. Plants can assimilate information, calculate outcomes, and respond using a complex series of molecular signaling pathways that are remarkably like those in our own brains. (Phillips 2002, 40)[8]

Much is made by Deckers of plants' sessility—of their purported inability to escape from danger. But regarding this as a disadvantage rather than an alternative adaptive advantage belies the fact that their sessility correlates with their greater genetic complexity with respect to animals. Plants must respond with far greater precision to the conditions they face than animals, since animals have the option to relocate (Trewavas 2009). Sessility thus requires an extensive, nuanced understanding of one's immediate environment not just to address dangers but also to access light and acquire water and nutrients. It is no wonder that plants have biochemical constitutions that permit them to identify specific insects from the taste of their saliva after being bitten, deter and poison predators, and even recruit insects to perform services for them (Buhner 2002, 162).

Consider these three examples: Tomatoes subject to damage by insects and herbivores produce methyl jasmonate as an alarm signal. Plants in the vicinity detect it and prepare for attack by producing chemicals that ward off their attackers (Farmer and Ryan 1990). They do not do so when subject to mechanically induced wounding, as from high winds. This indicates the capacity for discernment (Paré and Tumlinson 1999). Acacia trees excrete an unpalatable tannin to defend themselves when being eaten by animals. The scent of the tannin is picked up by other acacia trees, which then also excrete it (van Hoven 1991). And when attacked by caterpillars, some plants release chemical signals that attract wasps who attack the caterpillars (Thaler 1999).

Plants' sessility also accounts for their modularity; the operational control of growth and behavior is devolved among thousands of meristems at the tips of roots and shoots (Baluška, Volkmann, and Menzel 2005 and Brenner et al. 2006). This, too, has advantages unavailable to animals. Plants may not be able to escape danger, but they can survive extensive damage. According to Michael Pollan, their modular design permits plants to lose up to 90 percent of their bodies without dying.[9] Plants have no irreplaceable organs, so being food is not necessarily a death sentence. "There's nothing like that in the animal world. It creates resilience," Pollan states (2013, 92). Neither being sessile nor having a modular design entails sentience, of course. But they should not be taken as strikes against sentience either.

Among the core signatures of sentience is self-awareness: the capacity to distinguish self from other or, as DeGrazia puts it, "the ability to distinguish one's own body from the rest of the environment" (1996, 166). Available evidence suggests that this also is a capacity that plants have. Plants are capable of differentiating themselves from competing and noncompeting organisms and even discriminating between kin and nonkin (Karban and Shiojiri 2009). The former capacity is on display when plants must determine whether to compete for or share scarce resources with neighbors (Gersani et al. 2001, Maina et al. 2002, and Callaway et al. 2003). The latter capacity includes rejecting pollen contributed by plants with which the recipient shares an allele (Biedrzycki 2010).

Intelligence, Learning, and Intentionality

For sessile beings, the centralization of systems governing signal integration and response is a clear disadvantage. The decentralized character of these systems in plants is an extension of their modularity. Does this entail that plants lack the capacity for cognition? It certainly has been treated as a strike against them (James 1996, 166). But along with phenomenologists and researchers of swarm behavior, a number of cognitive scientists and philosophers of mind reject the proposition that not just cognition but intelligent behavior must result from centralized processes. Michael Marder suggests that intelligence "is not concentrated in a single organ but is a property of the entire living being" (2012b, 1370). According to Trewavas, "nervous systems are not necessary; complex networks are sufficient to create intelligent behavior" (2012, 773; see also Schull 1990, Warwick 2001, and Vertosick 2002). And Paco Calvo Garzón and Fred Keijzer (2009) contend that the "embodied cognition" of plants permits them to organize their behavior in complex and adaptive ways.

David Stenhouse (1974) suggests that intelligence should be conceptualized as the adaptively variable behavior of an individual organism during its lifetime. Stefano Mancuso defines the term "very simply" as the ability to solve problems. Plants clearly do both. Trewavas notes that "plants respond to an enormous variety of physical, chemical, and biological signals in the environment to maximize foraging for resources in two distinct but unpredictable environments: above and below ground" (2012, 773). This requires not just considerable physiological, morphological, and phenotypic plasticity but an acute awareness of one's lived conditions as well. It also requires the capacity to process information from both biotic and abiotic stimuli in order to discriminate between positive and negative experiences, make predictive assessments, weigh trade-offs, and formulate informed decisions (Hutchings and de Kroon 1994, de Kroon and Hutchings 1995, Grime and Mackey 2002, and Lyon 2006). Add to this the need to counter threats from predators and disease and the fact that "Mate selection is elaborate and underpinned by discriminating, complex conversations that precede and follow fertilization" (Trewavas 2012, 773) and it becomes clear that explaining plant behavior in terms of tropisms—reflex reactions to external stimuli—alone is simply inadequate.

Among others, Trewavas and Mancuso both contend that plant intelligence may be quite similar to swarm behavior, exhibited in the comportment of bird flocks, schools of fish, and insect and bee colonies. Swarm behavior is a manifestation of distributed intelligence, an emergent property of the organization of "mindless" individuals. Swarms are not coordinated by a central system. As Pollan states, "In a flock, each bird has only to follow a few simple rules, such as maintaining a prescribed distance from its neighbor, yet the collective effect of a great many birds executing a simple algorithm is a complex and supremely well-coordinated behavior" (2013, 92).[10] A similar phenomenon takes place within plants. Meristems act as do individual birds in a flock. They gather, assess, and respond to environmental data in local but coordinated ways that benefit the organism as a whole (Noble 2006, 113, Ciszak et al. 2012, and Marder 2013, 169).

While brains may provide a centralized site of cognition, they operate internally in much the same way. Here again is Trewavas:

> The hypothesis that intelligent behavior in plants may be an emergent property of cells exchanging signals in a network might sound

far-fetched, yet the way intelligence emerges from a network of neurons may not be very different. Most neuroscientists would agree that, while brains considered as a whole function as centralized command centers for most animals, within the brain there doesn't appear to be any command post, rather, one finds a leaderless network. That sense we get when we think about what might govern a plant—that there is no there there, no wizard behind the curtain pulling the levers—may apply equally well to our brains. (2005, 414)

Especially when we humans are young, perhaps the most salient aspect of our intelligence is our capacity to rapidly absorb vast quantities of information. Until we reach puberty, we are learning machines, as it were. We develop memories through our experience, and we establish and modify behavior and make predictions about future events in consonance with these memories. Like all cognition, our memories have a biochemical basis. This also is the case for plants, who remember and learn as well (Bhalla and Iyengar 1999 and Brenner et al. 2006).

Memory and learning in animals involves the development of new connections within the neural network, but this is not the only way that information is stored. Pollan remarks that immune cells remember their experience of pathogens, calling on these memories upon future encounters. So-called epigenetic effects from traumatic events can be passed to offspring, such as through genetic coding for high levels of cortisol (Molinier et al. 2006). These are the sorts of biochemical processes that contribute to the development of memories by plants (Sung and Amisino 2000 and 2004). Memories are thus inscribed in the bodies of plants rather than actualized in subjective consciousness, Marder asserts (2014, 33). Some plants learn faster than others, of course, but observed behaviors do not have the traits of innate or programmed responses. Plants even can be taught to learn more quickly and to better retain what they have learned (Gagliano et al. 2014).

With respect to making predictions, plants take great care to ensure that the energetic costs of possible actions do not exceed their anticipated benefits. Future sun and shade patterns are predicted based on the perception of reflected far-red light (Izaguiree et al. 2006). The Mayapple commits to a branching and flowering pattern early in its life, using an analysis of leaf and branch patterns above it to anticipate gaps and where light will land coming through them (Geber et al. 1997). The dodder, a parasitic and nonphotosynthetic plant, discriminates between potential hosts. Once a suitable host is chosen, it decides how many coils to make around it in order to optimize the return on energy investment (Kelly 1990). The stilt palm "walks" out of shade by differential growth of its roots (Trewavas 2005). And when fewer nutrients are found, plants are able to both accelerate root growth (Callaway 2002) and increase the rate of nutrient absorption (Jackson and Caldwell 1996).

I do not refer to plants making discriminating and careful choices flippantly. After watching time-lapse photography of a bean plant growing, Pollan describes a "seemingly conscious individual with intentions" (2013, 92). A bean sprout actively searches for something to climb, but it does not just grow any which way until it encounters something suitable. It seems to know exactly where to go long before making contact, perhaps through the use of echolocation or by sensing the reflection of light off its surroundings. Whatever the case may be, the bean sprout does not waste energy fruitlessly. "And it is striving (there is no other word for it) to get there: reaching, stretching, throwing itself over and over like a fly rod, extending itself a few more inches with every cast. [. . .] As soon as contact is made, the plant appears to relax; its clenched leaves begin to flutter mildly" (ibid.).

Perhaps Pollan simply witnessed the results of tropistic bio-chemical reactions. Biochemical reactions, yes, but our brains make decisions in the same way. Tropistic, no, or at least this is Marder's conclusion: "To be conscious is to intend something, that is to say to be directed toward the intended object. [. . .] In light of this definition, the intentionality of plants may be understood as the movement of growth, directed toward opti-mal patches of nutrient-rich soil and sources of light" (2012b, 1367; see also Darwin 1880, 573). It involves the deliberate prioritization of some sources over others. Animals' inten-tions, the enactment of directedness, are carried out via muscle movement. Intentionality in plants "is expressed in modular growth and phenotypic plasticity" (ibid.; see also Hutchings and de Kroon 1994). Again, it pays in terms of survivability for sessile beings to be highly attentive to the context of growth and development. Vigilance and the capacity to respond as one's context changes are key. Sometimes this requires non-cognitive action, something akin to our instinctive drive to flee. Sometimes careful and focused cognition is in order. It would seem that plants do both, even if these actions generally occur in a time frame that we have difficulty discerning.

But why should the time frame in which plants operate have much of an effect on our judgment of their capacities anyway? Imagine the following, Mancuso states: "A race of aliens living in a radically sped-up dimension of time arrive on Earth and, unable to detect any movement in humans, come to the logical conclusion that we are 'inert material' with which they may do as they please. The aliens proceed ruthlessly to exploit us" (quoted in Pollan 2013, 92). We are not inert material, of course. Neither are plants. If we judge that it is wrong for aliens of the sort Mancuso describes to do as they please with us, then we must accept that it is wrong for us to do the same with plants.

Teaching and Nurturing

Keystone species—ironwoods, for example—are usually large, elderly trees who contain more mass than any other plant in a given ecosystem. Each individual keystone forms an archipelago around it that contains scores of associated plants. The formation of archipelagos takes centuries. How and why they form as they do is not well understood. As Stephen Buhner states, "Quite often, before it can establish itself in a new location, a keystone species must have a plant that goes first and prepares the way. These initial species, usually selected from among the community of plants that grow with the keystone plants, are the outriders, the plants whose emergence signals the movement of plant species in mass. [. . .] These plants move first and essentially determine what keystone plants will grow where and when. In this way, they act as 'filters' through which keystone species are sifted" (2002, 181). Outriders make note of the quality of the soil, water, and light and send a return signal to tell the keystone species where and when to send seeds. Neither wind nor animal dispersal can explain how these seeds move where they move, Buhner contends. "The distances are too far, the dispersal patterns too unusual. But by whatever means, the seeds answer the chemical call sent by the nurse plants" (ibid.).

Once established at the new location, keystone species then call to them soil bacteria, mycelia, and the plants that constituted their previous archipelago. "As the plants arrive, the keystone chemistries literally inform and shape their community structure and behavior. This capacity of keystone species to 'teach' their plant communities how to act was widely recognized by indigenous and folk taxonomies" (ibid., 183).

Plants also can nurture other plants directly. In the Great Basin of Utah and Nevada, sagebrush nurses piñon pines until the pines are old enough to grow on their own. The sagebrush alters the soil chemistry and provides physical protection from the elements (Callaway 1995).

Plant Sentience: The Speculative Leap

So plants have an information-processing and response system that is homologous to a central nervous system, and they exhibit some key characteristics of beings who suffer. Evidence supports the proposition that they are self-aware and highly attentive to their environments; exhibit intelligence and intentionality; and can remember, nurture, learn, and even teach. Perhaps I have succeeded in raising further doubts about the sentientist stance regarding plants. Perhaps not. Appealing to the parsimony principle, "which is the cornerstone of scientific approaches," Paul Struik and his colleagues assert that the kind of evidence I catalog does not permit "unfounded philosophical speculation" (2008, 369). From their perspective, I do precisely this by bringing plant and animal behaviors under the same semantic umbrella. This is an informed choice on my part, but it surely involves speculation as well.

Yet whether or not my speculation violates the parsimony principle is itself a philosophical question, as is the merit of adhering to this principle in the first place. Inferring the sentience of other-than-human animals requires empirical evidence on a par with what I here provide regarding plants. But neither determination can rely on empirical evidence alone. Matthew Hall rightly remarks that "our perception of plants depends heavily on our philosophical orientation" (2011, 35). But at the end of the day, the need to take a speculative leap is a professional hazard that every philosopher must face. Marder infers plant sentience "from the fact that plants explore and pursue unevenly distributed resource gradients, assess environmental dangers from biotic and abiotic stressors, and gather and constantly update various types of information about their surroundings" (2012b, 1368). But do they *actually* explore, assess, gather, and update? These are signs of his leap, and I see no compelling reason to reject it. For what it is worth, plants make up 99 percent of the biomass on Earth. They dominate every terrestrial environment. We animals are the outliers. Of course, it is possible that only we animals—indeed, just a subcategory of us—are sentient, but a shift in the burden of proof is overdue.

THE EXPANSIONARY SENTIENTIST CASE FOR VEGETARIANISM

Yet consider the following from Singer:

> Assume that, improbable as it seems, researchers do turn up evidence suggesting that plants feel pain. It would still not follow that we may as well eat what we have always eaten [i.e., an omnivore diet]. If we must inflict pain or starve, we would then have to choose the lesser evil. Presumably it would still be true that plants suffer less than animals, and therefore it would still be better to eat plants than to eat animals. Indeed, this conclusion would follow even if plants were as sensitive as animals, since the inefficiency of meat production means that those who eat meat are responsible for the indirect destruction of at least ten times as many plants as are vegetarians! (Singer 2002, 236)

Yes, plants may suffer. Their suffering may equal that of animals. But Singer's appeal to the inefficiency of omnivorism is powerful. It is widely accepted that far fewer sentient beings are harmed by vegetarianism than by omnivorism (Regan 1983, 305, Pringle 1989, 28, and Gruzalski 2004, 134). This would seem to be sufficient simply to reaffirm sentientism.

If we add to this the ecological harm done to *all* living beings by omnivorism as this diet is industrially supported today, the case for reaffirming sentientism would seem to be even stronger (Wenz 1984). Deforestation, soil erosion, loss of soil fertility, and desertification all result directly from the production of feed and from overgrazing (Taylor 1986, Durning and Brough 1991, Rifkin 1992, 201, and Pluhar 2004). Considering that well over half of the total land area in the continental United States is used to produce meat and dairy products (Hill 1996, 106), the widespread persistence of industrially supported omnivorism is ecologically catastrophic. Meat production is the single largest source of water usage in the United States and by far the greatest cause—as much as 80 percent—of water pollution worldwide (Schleifer 1985, 68, Amato and Partridge 1989, 19, and Hill 1996, 111). Thousands of liters of water are needed to produce a single kilogram of American beef, according to some estimates, and cattle feedlots account for more than half of toxic organic pollutants found in fresh water (Rifkin 1992, 218ff.). Factory farming is also a leading cause of biodiversity loss and global climate change (Regenstein 1985, Durning and Brough 1991, and Fox 2000). Vegetarianism is not necessarily ecofriendly of course, just as more ecofriendly forms of meat production exist. On balance, though, the evidence in favor of the ecological benefits of vegetarianism appears to be overwhelming.

But simply countenancing killing plants instead of animals is unacceptable so long as we intend to adhere to the principle of nonmaleficence. If sentientists refuse to be implicated in killing any sentient being for food, then perhaps they are committed to nothing less than conscientious suicide after all! Maybe instead they must explore the prospects of inedia, or breatharianism, according to which it is possible to live without consuming food. But we can take both options off the table. "Ethics is intended as action-guiding within the framework of ongoing human life," Gruzalski notes (2004, 134). The satisfaction of our basic needs may harm other sentient beings. We must be permitted to do what is necessary to survive so long as we act in ways that cause the least possible harm. We have already seen that this is consistent with the principle of nonmaleficence.

But reaffirming sentientism—by accepting at face value the position that Singer outlines in the passage at the beginning of this section—does not commit us to doing the least harm. As I hope the case I now make for expansionary sentientism shows, how plants are treated in the process of cultivating and gathering them for food matters morally. This means that not all ways of practicing vegetarianism are equal. Some may cause more harm than others. If we wish to adhere to the core principles of moral standing, we can do better than adhering to the dictates of sentientism.

For the expansionary sentientist, sentience remains a sufficient condition for being granted equal moral standing. But killing and eating beings with equal moral standing is permissible so long as great care is taken to adhere to the principle of nonmaleficence. Aside from the consumption of roadkill, eating animals requires killing or maiming them. This is not necessarily the case with plants if steps are taken to facilitate their well-being. This gives us presumptive reasons to continue to favor vegetarianism while also giving plants their moral due.

Indeed, the expansionary sentientist endorses a fourth principle of moral standing, the *principle of relatedness* (Detwiler 1992, 238f., Rose 1992, 187, and Sanchez 1999, 207ff.). This principle acknowledges the abiding connection we have with our distant green relatives. Yes, we share sentience, but our connection goes deeper than this. Whether directly or indirectly, we depend on plants for our subsistence (Hall 2011, 89). To the extent that we are what—or who—we eat, we *are* plant matter. They make our lives, hence our *moral* lives, possible. "Thus, the resources of the earth do not belong to humankind;

rather, humans belong to the earth," asserts Manuka Henare (2001, 201). This proposition provides sufficient grounds to conclude that our relationship with plants is heterarchical. We are moral equals. Presuming otherwise is "kingdomist" (Quinn 1994, 165) in the same way that positing a morally relevant distinction between humans and other animals is speciesist. Life, shall we say, is life.

In consonance with the principle of relatedness, we should view the cultivation of plants for consumption as a responsive, collaborative, reciprocal, and caring process. According to Hall, our partnership with plants must be "based on respect for the awareness and autonomy of the cultivated" (2011, 35; see also Kohák 1997). This includes "nurturing all the potentialities proper to [plants], including those unproductive from the human point of view," states Marder (2012a). These steps mitigate and perhaps even offset the harm done by taking their lives when we have no other viable choice.

But taking their lives can be kept to a minimum, too, since expansionary sentientism accommodates a key aspect of plant modularity. Plants are anatomically constructed such that fruit (and vegetables, pulses, nuts, and so on) and leaves can serve as food without placing the life of the plant at risk. Indeed, under given circumstances, fruit and leaves can be harvested for consumption without causing plants any harm at all (McCabe 2007, 175, Gollner 2008, and Curnutt 2011, 367). Flowers, fruit, and leaves are attached to plants by means of receptacles, which are modified or expanded portions of the stem specifically designed for abscission. Abscission, the detachment of fruit and leaves from the main body of plants, is an important part of their developmental process. The connection between the receptacle and the flower, fruit, and leaf is structurally weak in comparison to the rest of the plant. Physiological processes further weaken the connection upon ripeness, maturity, and autumnal foliation. Both perennials and annuals thus drop flowers, fruit, and leaves of their own accord (Roberts et al. 2002). This contributes to seed dispersal, prevents the spread of disease, and optimizes energy expenditure. So abscission plays a critical role in plants' health and reproductive success (Webster 1968 and Sexton 1995).

Perhaps Mancuso has a point, then, in declaring that "Plants evolved to be eaten—it is part of their evolutionary strategy" (quoted in Pollan 2013, 92). But granting plants equal moral standing entails that moral agents must account for optimal conditions for abscission and other bases for their flourishing when using them for food. As far as is possible, plants should be permitted to live out the complete life cycle that their climate and location permits. Fruit and leaves can be picked, preferably upon maturity, but the plant should be left alive and intact. Eating root vegetables requires killing the plant, since it must be pulled from the ground. But root vegetables mature at the end of the plant's life cycle, which is why many are fit for consumption in autumn. So they also can be harvested when the time is right. Some can even regenerate—yams, potatoes—if part of the root is cut off and replanted (Fox 1999, 155). And we should not overlook the importance of seed dispersal. Like other animals, we can contribute our excrement, our "night soil," to the land. "That's how China kept growing food on the same fields for millennia," remarks Bill McKibben (2010, 165).

These considerations would seem to justify fruitarianism, since fruitarians commonly advocate eating only what falls or would fall of its own accord from a plant (Samour 2005, 14, McCabe 2007, 175, and Gollner 2008). Most fruitarians consume pulses: beans, peas, and other legumes. Others likewise eat raw fruits, dried fruits, nuts, and cold-pressed oils. Still others eat seeds, although this is a point of contention. Seeds are the basis of future plants, so eating them represents a denial of propagation (Lovewisdom 2005). Some countenance consuming cooked food, while others maintain a fully raw diet.

It is not entirely clear, however, that fruitarianism is a viable diet long term. Evidence supports the proposition that it can lead to deficiencies in calcium, protein, iron, zinc, vitamin D, vitamin B, and essential fatty acids. Many experts suggest that adults should not adhere strictly to fruitarianism for too long. Moreover, they are of one voice that it is wholly inappropriate for children, as it may put their growth and development at risk (Ensminger 1993, 1876, and Holden and MacDonald 2000). Perhaps this is not so. I am in no position to get involved in the finer points of whatever debate there may be. It should go without saying, though, that if expansionary sentientists reject principled suicide, they reject nonviable diets as well. And if fruitarianism is not viable, then we must establish criteria according to which it is acceptable to kill and eat plants.[11]

Bringing an end to conventional agriculture, or at least doing what we can to avoid eating its products, surely is in order. The optimal source of food for plants is nutrient-rich soil, not the petrochemicals on which conventional agriculture depends. Organic farming, preferably done locally and on a small scale, is the best choice if we are to give due consideration to the needs and interests of plants. Each individual plant can be given attention, composting and crop rotation keep the soil healthy, and concerns about industrialized genetic modification are avoided (Cummins and Lilliston 2000 and Shiva 2004). Marder (2012a) asserts that it is particularly harmful to grow plants from sterile seeds, a practice that is now common because it requires farmers to purchase new seeds for each growing season, for it robs plants of their reproductive potential.

Critics of organic agriculture argue that it produces lower yields, so it requires more land use to produce the same amount of food as conventional farms. Replacing conventional agriculture with organic agriculture hereby would result in more widespread deforestation and biodiversity loss (Trewavas 2001 and Connor 2008). Even if organic agriculture is beneficial to the plants being cultivated, the benefit is more than offset by the damage done to the land and accompanying biota.

There are three problems with this argument. First, the concerns that motivate it reflect a canard. Those who oppose organic agriculture on these grounds give little consideration to the fact that our species faces serious problems associated with overpopulation. Our ranks have grown exponentially in the past several decades due almost entirely to our singular reliance on the abundance of energy from the "ancient sunlight" contained in fossil fuels (Hartmann 2013, 8). So opponents of organic agriculture express concern about whether moving away from conventional agriculture can feed a growing human population when the more apt concern should be with the ecological limits of our population growth. I will explore this matter in more detail later.

Second, opponents of organic agriculture ignore the amount of food that goes to waste—some 50 percent of produce in the industrialized world—while nearly one billion people starve. So they overlook the extent to which starvation is a political problem associated with how food is distributed rather than a result of food availability decline (Sen 1981). Namely, food is accessible to those with the money to pay for it, while the poor starve even though adequate nourishment is within relatively easy reach. Even if complete reliance on organic agriculture would lead to the production of less food, this may not be the problem its detractors suggest it is.

Third, the evidence in favor of the proposition that organic agriculture does more harm than good is weak. When energy inputs and externalities associated with environmental degradation, adverse health, and the hollowing out of rural communities associated with conventional agriculture are accounted for, organic agriculture measures up exceptionally well (Duram 2005, 15f., and Shiva 2012). Even when these factors are not considered, with good practices, appropriate choice of crop types, and adequate growing conditions, organic yields may be able to match conventional yields (MacIntyre 2009, de Schutter 2010, and Seufert, Ramankutty, and Foley 2012). And over time, soil erosion caused by conventional agriculture reduces its yields in the absence of more intensive petrochemical use (McKibben 2010, 172). This eventuality also adversely affects the nutritional content of conventionally grown foods, a clear indication that plants grown in this way do not enjoy conditions of optimal flourishing (Brandt and Mølgaard 2001, Biao et al. 2003, Thomas 2003, and Crowder et al. 2010).

Consider, for example, the advantages of intercropping: the cultivation of two or more crops simultaneously on the same field. Intercropping has a number of benefits for plants. It makes for more efficient land use. It increases resistance to disease. It introduces semiochemicals that better protect cultivated crops from harmful insects without the use of pesticides, in part by increasing the populations of these insects' predators. It increases soil microorganism activity, which protects the topsoil and supports biodiversity. And it enhances weed suppression, which means that far fewer inedible plants must be killed (Liebman and Dyck 1993 and Altieri 1995). All these factors improve the health and well-being of cultivated plants, as the productivity of farms that employ intercropping makes evident (Ouma and Jeruto 2010). Perhaps this also is partially the result of the fact that farmers who employ intercropping have a better sense of their ecosystem and its microclimates (McKibben 2010, 167). So they likely have a much closer relationship with the beings in their care than do conventional farmers.

Expansionary sentientists nevertheless acknowledge the inevitability of what Hall calls "colliding agendas" (2011, 163) between plants and us. Barring new evidence that supports the viability of fruitarianism, it is a "messy fact of life" that no matter how caring we are toward our distant green relatives, our survival sometimes requires that we "violate plants' integrity, curtail their flourishing, and ultimately harm them. Human people must sometimes act against the interests of herbs, shrubs, and trees that are actively striving to live and reproduce" (ibid., 111). Unlike sentientists, expansionary sentientists exhibit no compunction to ignore the intimacy of our relatedness to plants or deemphasize plants' capacities "in order that human beings might pretend that their lives can operate without harming the integrity of other beings" (ibid.). Acknowledging this does not give moral agents license to harm plants unnecessarily or kill them indiscriminately. Instead, it should make us receptive to the fact that killing is part of living (Jensen 2000, 175); nutritional exchange is a necessary condition for ecological health. Each and every living being is a natural-born killer *and* natural-born food. Going a step further than Mancuso, we *all* evolved to be eaten as well as to eat. It can be no other way.[12]

The expansionary sentientist thus embraces a form of vegetarianism that emphasizes respect for and responsibility to our
distant green relatives. The moral defense of vegetarianism on
expansionary sentientist grounds remains true to the principles of
moral standing without turning a blind eye to the fact that there
is no complete escape from doing harm when it comes to eating.
Again, it can be no other way. Even moral eaters are messy eaters.
A tree who spoke to Jensen put it particularly well: "You're an
animal, you consume, get over it" (2006, 283).

CHAPTER 3

ANIMISM

I have argued thus far that expansionary sentientists provide a more satisfactory defense of vegetarianism than do sentientists. Expansionary sentientists account for the ample scientific evidence that plants are sentient. Because sentientists overlook or conveniently ignore this evidence, they turn a blind eye to ways in which plants can be harmed that are comparable to how animals can be harmed. That plants are sentient does not rule out vegetarianism. It is still preferable to omnivorism, according to expansionary sentientists. But killing and eating plants requires that they be treated with care and respect when cultivated and harvested. So long as expansionary sentientists follow through on these dictates, they better adhere to the core principles of moral standing than do sentientists. Simply put, they do less harm.

I worry, though, that expansionary sentientism retains troubling vestiges of sentientism that may diminish or thwart our capacity to be as responsive to the needs and interests of plants, and perhaps even other-than-human animals, as we should be. Retention of these vestiges may lead expansionary sentientists to harbor, if unwittingly, a privileged view of humans with respect to the more-than-human world. So while expansionary sentientism provides a better basis to defend vegetarianism than does sentientism, both the former and the latter presuppose a conceptual framework that may permit slipping into problematic ways of thinking.

Consider the following: Expansionary sentientism is an extensionist ethic. It is emblematic of Singer's (2011) call to *expand the circle* of who is worthy of moral standing. This is hardly damning on its face, but consider what the idea of expanding the circle implies. We humans are at the circle's center, according to Singer. It is our job, our duty, to confer moral standing. This presents two problems.

First, Val Plumwood worries that the idea of expanding the circle quite literally "treats others of value just to the extent that they resemble the human as hegemonic center, rather than as an independent center with potential needs, excellences, and claims to flourish of their own" (2002, 167). Other-than-human beings are worthy of moral standing, yes. But they are worthy of moral standing because they have characteristics sufficiently like ours. In Chapter 2, I noted that expansionary sentientists contend that we should be aware of what is important to plants quite independently of what is important to humans. My concern is that expansionary sentientists nevertheless are still involved, again with reference to Plumwood, in the "quest to discover which parts of nature are sufficiently 'well qualified,' usually by being proved to be enough like us humans, to deserve some sort of extension (the leftovers) of our own ample feast of self-regard" (ibid., 168). Given their focus, the needs and interests of other-than-human beings that happen most closely to resemble what are regarded as quintessentially human needs and interests may end up being given greater moral weight than those that do not. As a result, expansionary sentientists' attempts to adhere to the principle of equal consideration break down.

Second, if we do adhere to the principle of equal consideration to the letter, we should reject that we have the right, let alone the duty, to confer moral standing in the first place. "Humans don't even have the moral authority to extend ethics to the land community," proclaims Dennis Martinez. "We don't have the right to extend anything. What we have the right to do is to make our case, as human beings, to the natural world" (Martinez, Salmón, and Nelson 2008, 89). This is what it means to live in a world, to be part of a community of life, marked by mutuality and reciprocity. Surely, this represents our best vision of caring about and respecting others as autonomous beings quite independently of how they measure up to us.

So expansionary sentientism presupposes a conceptual framework that, by all appearances, threatens to invalidate the principle of equal consideration. This in turn leaves expansionary sentientists prone to violating the other principles of moral standing as well—the precautionary principle and the principles of nonmaleficence and relatedness. Wittingly or not, like sentientism, expansionary sentientism can lead us to harm others by overlooking needs and interests that diverge too greatly from ours. And this damages our connection to them and theirs to us. This surely does not provide the best basis for a moral defense of vegetarianism. Perhaps we should try again by first identifying a better conceptual framework than sentientism provides and then seeing where it leads in terms of developing a worthy defense.

To set the stage, I begin by laying out Plumwood's critique of what she calls "ontological veganism," which shares some of the basic problems exhibited by the sentientist conceptual framework. I then outline Plumwood's alternative framework, which is intended to better account for "ecological complexity or the continuity of life" (2000, 302) than does ontological veganism. Specifically, Plumwood develops a conceptual framework that highlights humans' ecological embeddedness and animality. This shifts our focus away from identifying purportedly quintessentially human characteristics that other beings happen to share to acknowledging our indistinctness from the rest of the community of life. In the process, we are able to shed the residual prejudice—or as much of it as inevitable cognitive biases permit (Ariely 2010)—that expansionary sentientists ultimately fail to escape.

While Plumwood advocates attending to the needs and interests of plants, her main focus is on ecological relations between humans and other animals. In order to take even fuller account of the complexity and continuity of life than she does, I rely instead on an *animistic* conceptual framework.[1] Animists maintain that the world is full of people, only some of whom are human or perhaps even sentient in the typical way that we understand this concept. The other-than-human people who populate our world do not need us to grant them moral standing. Martinez is right that we do not have the moral authority to do so anyway. But they do deserve for us to cultivate attunement to their means of communication. We ought to listen to them and take seriously what they have to say. By doing so, I argue, we should recognize that what I call a *care-sensitive ecological contextualist* defense of vegetarianism is preferable to the expansionary sentientist defense. According to care-sensitive ecological contextualism, vegetarianism is not universally justifiable, but it is justifiable in certain contexts—most notably including urban settings. And whether or not we practice vegetarianism, we must make every effort to be attuned to, responsible for, and responsive to the needs and interests of those whom we make our food.

Ontological Veganism and the Residue of Human Mastery

According to Plumwood, ontological veganism "is a theory that advocates universal abstention from all use of animals as the only real alternative to mastery and the leading means of defending animals against its wrongs" (2012, 78). Breaking this down, we can begin by noting that ontological *vegetarians* reject regarding animals as edible. To see animals potentially as food—specifically as nothing but food—connotes that their existence, or their ontological status, amounts to being means exclusively to our ends. And this suggests, quite problematically from the viewpoint of ontological vegetarians, that they owe their lives to the fact that they are edible for us.

Ontological *vegans* take the further step of rejecting any use of animals whatsoever (Plumwood 2000, 297). Carol Adams serves as an exemplar of ontological veganism, Plumwood contends. Adams emphasizes that humans do not need to prey on other animals to survive. Indeed, there is nothing the least bit natural about the systematic torture and slaughter of billions of beings via "the grossly inhumane institutions of the animal-industrial complex" (1993, 200), Adams proclaims. No reform of slaughterhouse conditions can possibly right the wrongs done to the animals who meet their end in them. So we must resist not just an "ontology of animals as consumable," states Adams, but any attempt to treat "animals as usable" (ibid., 205). This is the only practicable way to reject the presupposition of human mastery and the systematic exploitation of animals that it supports.

According to Adams, to make any mention of the proposition, as Nel Noddings puts it, that "it is the fate of every living being to be eaten" (1991, 420) implies that inevitable processes of death and decay are somehow comparable to the practices of factory farming. But something is clearly wrong here, Plumwood asserts. Adams mischaracterizes Noddings's claim and then employs this mischaracterization to assert that the use of animals is invariably exploitative. Indeed, Adams maintains that to be used by another for any purpose whatsoever is to become nothing but an instrument for the fulfillment of their desires. There is no middle ground. Either one is entirely free from others' use or one is their instrument. Perhaps Adams is wary about exhibiting any nuance that proponents of factory farming can misapply. But this should not spare her from criticism, Plumwood declares, particularly since she ends up tacitly supporting human mastery herself.

How is this so? Curiously, Plumwood contends, the view that to be edible is to be denied moral standing has its roots in "the taboo of envisaging the human in edible terms" (2004, 349). This taboo serves as a cornerstone for the defense of human mastery. Ontological vegans (and ontological vegetarians) simply extend it to encompass animals as well. So animals essentially become honorary masters alongside humans. Ontological vegans rightly reject the "radical separation between humans and other animals," Plumwood notes (ibid.). But they still buy into a view, shared by proponents of human mastery, that does not square with the reality of our ecological embeddedness. We are all inescapably natural-born killers and natural-born food, I stated in the previous chapter. Noddings is right. Everyone survives on the lives of others and fosters others' lives in death. We are all users and are used in turn, and using and being used need not involve exploitation.[2] To regard this as taboo is simply to deny that which is unavoidable.

Ontological vegans also are correct to challenge the consignment of animals to the status of mere resources. But they do so in the wrong way. Instead of acknowledging the ecological embeddedness of all members of the community of life—specifically, that all are part of the food web—they extend to animals membership in an ecologically empty category. There is no master position outside the food web. To assume otherwise encourages ontological vegans to leave unchallenged the proposition that there must be beings with no moral standing if we are to eat. From this perspective, "to be food is to be degraded," Plumwood proclaims (2000, 302; see also Keith 2009, 85f.). Edibility requires moral exclusion.[3] Other social phenomena surely reinforce this notion, most notably our stark separation today from the processes that produce our food. But this makes the position maintained by ontological vegans no less problematic.

Consider what this means for plants. According to Plumwood, Adams "simply ridicules the idea that plant lives could have individual value or that plants could have forms of awareness, for example, in exactly the same way the idea that animals could have minds was ridiculed earlier by those who urged that animals were beneath consideration" (ibid., 301). But in this case, Adams's position is more nuanced than Plumwood suggests. Adams challenges defenders of factory farming who propose that vegetarians are hypocritical to reject killing and eating animals while condoning killing and eating plants. If plants do not count morally, they argue, neither should animals. She responds by seeking to highlight why morally distinguishing plants and animals is justifiable—why plants are not deserving of the moral standing that animals should enjoy.

But we can acknowledge that plants have not just moral stand-ing but equal moral standing *and also* reject factory farming. We do not need to denigrate plants to hold that the systematic tor-ture and killing of billions of animals in slaughterhouses every year is morally abhorrent. Adams does not reject alleviating the ill use of plants. She echoes Barbara McClintock by advocating that we engage with plants with a "feeling for the organism" (Adams 1994, 107; see also Keller 1983 and Hogan 1995, 48). And Adams is sympathetic to the idea that plants are misused by the factory-farming industry since they are treated with petro-chemicals and genetically modified in the process of becoming animal feed, as is the fate of much of the plant life grown via conventional agriculture. But this does not prevent her from des-ignating plants as "intrinsically renewable resources" (ibid., 106) because they "can regenerate themselves" (ibid., 222, n. 61). Resources. Mere means. And she regards as a distortion of "the lived reality we know as this world" and "a failure to participate in embodied knowledge" that carrots can be exploited in ways that are comparable to how horses, pigs, or chickens can be exploited (ibid., 107).

While what Adams suggests may have intuitive appeal, I believe that it is best to withhold support for one simple reason. I am not convinced that *her* intuitive sense of what constitutes lived reality is consonant with "the" lived reality. I have strong suspicions that what she takes to be representative of reality may be distorted in ways that most people of our culture find hard to recognize.[4]

FROM ONTOLOGICAL VEGANISM TO ECOLOGICAL ANIMALISM

Ontological vegans deserve credit for highlighting the horrors faced by the animals who live and die via factory farming. I also appreciate Adams's worries about providing fodder for apolo-gists of the animal-industrial complex. But we can adamantly oppose such practices in a way that "more thoroughly dis-rupts the ideology of mastery," Plumwood asserts (2012, 78). For whereas ontological vegans focus solely on "situating non-human [animal] life in ethical terms," ecological animalists also attend to "situating human life in ecological terms" (2004, 347), which entails highlighting our own animality. Taking these steps is necessary if we are fully to adhere to the core principles of moral standing. This becomes clearer by examining a key practice advocated by Plumwood: sacred eating.

Sacred Eating

All of us live and die within ecological systems of mutual use via food and energy exchange. We are "akin to rather than superior to" all other members of the community of life, Plumwood states (ibid., 346). There is no "outside nature," no ontological position of privilege, no ecological separation between humans and other-than-human beings. Acceptance of this proposition by ecological animalists marks a stark point of contrast with ontological vegans.

With this in mind, state ecological animalists, imagine a world in which humans (although perhaps not you or I personally) eat other animals without denying or overlooking that they, like we, are more than food. Imagine not just how this would change our relationship with them but how it would change us. To simultaneously be food and more than food is to be a *who* rather than a *what*, a locus of agency, a potential negotiating partner. This view "affirms an ecological universe of mutual use, and sees humans and [other] animals as mutually available for respectful use in conditions of equality," Plumwood remarks (2012, 78). One can be ontologized as edible—one can be viewed as food—according to ecological animalists without being treated as inferior. Humans can be predators, seekers of food and dealers of death for both other animals *and* plants, without seeking mastery. Yes, you read this correctly. Understood in general terms, to be a predator is to prey on other living beings. To be part of the community of life is to be both predator and prey (Keith 2009, 56). Specifically, we can take the lives of animals seriously "in both individual and ethical terms" (Plumwood 2000, 289) without being forced to deny that we are all natural-born killers and natural-born food. This is the essence of what Plumwood calls "sacred eating," which is no utopian view. Indeed, it is anything but utopian, as I will discuss momentarily.

Sacred eating operates on the model of a gift exchange, Plumwood asserts. It is "a sacrament of sharing and exchange of life in which all ultimately participate as food for others" (ibid., 299). We cannot embalm the dead thoroughly enough or entomb them sufficiently tightly to stave off this eventuality. All these practices do is make the dead a more toxic source of food (Jensen and McBay 2009, 139ff.).

The good human life surely involves reducing our toxicity, although this is an uphill battle to the extent that we live in a world awash in chemical pollutants (Williams 2012). Jack Forbes contends that living a good life also entails acknowledging that because others must die for us to live, our unavoidable complicity in killing "requires spiritual preparation. Moreover, one should feel the pain and sorrow of killing a brother or sister, whether it is a weed, a tree, or a deer. If one does not feel pain, one has become brutalized and 'sick'" (2008, 14). Living at a distance from the killing done for us to have food is no excuse for failing in this regard, even if it makes the task of attunement that much more difficult.

The Relational Hunt

A number of environmentalists, most notably José Ortega y Gasset (1985) and Holmes Rolston (1988 and 1989), suggest that if we are to regard eating as sacred in the way here described, then we should regard certain hunting practices as sacred, too. They defend what Adams calls the "relational hunt." According to proponents of the relational hunt, how an animal is killed for food affects whether or not the death and consumption of that animal is morally justifiable. If the approach to hunting and killing implied by sacred eating is applied to hunting, animals are not ill-used. As Adams states, "Instead, it is argued, this method of killing animals [. . .] reflects reciprocity between humans and hunted animals" (1994, 103).

Unsurprisingly, Adams is not convinced. Employing a different method for killing other-than-human animals still requires violence, she contends, and the end result is the same: a corpse. Indeed, it is not at all clear how the relational hunt involves a reciprocal exchange. Does hunting and killing with a specific mentality provide a gift on par with the animal "giving" its life? Does the honor of being a sacrifice, a sacred offering, really provide recompense for ending up a corpse? Surely not, Adams concludes. This marks a clear rejection, by her at least, of this aspect of ecological animalism.

Although Plumwood contends that Adams is making a universal claim against hunting, Adams is quite clear that her focus is squarely on how the people of our culture exploit and distort indigenous practices. She is not condemning the diet of the Inuit, for example, who rely heavily for food on seals, sea lions, and caribou. Nor does she care to weigh in on claims to fishing or whaling rights by indigenous peoples (see ibid., 83; see also Kheel 1995). But she does wonder why environmentalists like Ortega y Gasset and Rolston do not "hold up as a counterexample to ecocidal culture gatherer societies that demonstrate humans can live without depending on animals' bodies for food" (ibid., 105; see also Smith 1993).

Rod Preece, a fellow adherent of ontological veganism, takes an even stronger stance against the relational hunt, challenging a common defense of sacred eating offered by indigenous hunters. Karen Warren (1990, 146) quotes a Lakota elder who is instructing his son about how to engage with deer during the hunt. Shoot "your four-legged brother" in the hindquarters to slow him down without killing him, the father advises. When he has fallen, look into his eyes, which convey his suffering. Take in his experience; feel his pain. Then use a sharp knife to cut his throat so that he dies quickly. As you do so, ask your brother for forgiveness. Offer prayers of thanks for his sacrifice and promise to return to the earth when you die so that you can nourish the land, the plants, and ultimately other four-legged brethren.

Preece voices a number of concerns. Here are three. First, the approach to hunting and killing advocated by the Lakota elder "not only increases the animal's pain and suffering, but the hunter must be well aware of the increased harm he is committing" (2004, 238). It is hardly necessary to wound and terrorize one's prey. This seems like an obvious violation of the proscription against unnecessary harm. Why not instead dispatch with his life as quickly as possible? Second, it is deeply misleading to depict one's prey as a brother. No one in their right mind would engage with their immediate relations as the hunter does with the deer. Third, it is contradictory to ask one's prey for forgiveness for killing him if his life truly is given as a gift. By asking for forgiveness, hunters must sense that they are violating a relationship rather than affirming it. As a result, Preece concludes, sacred eating is a morally dubious practice—and ecological animalism is a morally dubious guide for action—even when it is carried out in the way the Lakota elder describes.

Although Preece's criticisms do not nullify the wider merits of sacred eating, his concerns about these specific practices warrant attention. I do not know how the Lakota elder would respond. I suspect that he would request that we be open to the proposition that these practices represent a longstanding agreement between hunters and hunted that has played an integral role in facilitating the health and well-being of both parties. But I leave this matter aside for the time being, since my quarry lies elsewhere. Recall that Adams questions why environmentalists appeal to the practices of indigenous hunters rather than to practices of indigenous gatherers. Taking the latter approach, she contends, would more thoroughly disrupt the norms that condone domination of the more-than-human world by the people of our ecocidal culture. Moreover, she argues elsewhere that we should not overlook that eating a vegetarian diet can be construed as "feeding on grace" (2001), or feeding oneself in a way that conforms to divine intentions. Refusing to eat animal flesh and instead celebrating the consumption of rice, for example, arguably is worthy of being considered a hallowed practice (1990, 143ff.).

For his part, Preece wonders whether Lakota hunters give due consideration to sentience, since they equate killing the brother deer with killing the sister flower (2004, 238). Furthermore, since hunters like the Lakota elder acknowledge that "respectful killing is preferable to disrespectful killing," it is curious that they do not conclude that "it would be better if there were no killing at all, if no harm were inflicted. Yet, of course, no Native nation has ever drawn the practical implications of that conclusion" (ibid., 239).

Note the general thrust of both Adams's and Preece's three conclusions, all of which represent tenuous bases for their rejection of aspects of ecological animalism: First, we can feed on grace by abstaining from eating animals. Second, we fail to give proper weight to sentience by equating animals with plants. And third, we can abstain from killing by rejecting the hunt in favor of eating plant life. The first conclusion reflects a clear denial of the moral standing of plants. The second defies the evidence in favor of plant sentience and, in the process, reflects the prejudice of human mastery exemplified by both the expansionary sentientist and the ontological vegan. And the third is an ecological fallacy, plain and simple. We began to see why in the previous chapter and gained a bit more clarity by assessing the basic tenets of ecological animalism. But this will become much more obvious as we proceed. Indeed, why each of these claims is so problematic becomes quite a bit clearer when regarded through an animistic lens.

From Ecological Animalism to Animism

Do I exhibit a degree of hubris by claiming to be able to explore the animist's world with any sort of credibility? I am no anthropologist; I am not professionally trained to speak authoritatively about others' cultures. Moreover, animism is not part of my everyday lived experience. So I face considerable risk in speaking as both an amateur and an outsider. This is obviously a dangerous prospect. It is why I rely quite heavily on the firsthand accounts of animists and the considerations of anthropologists who catalog their ideas.

But perhaps I overstate my unfamiliarity with animism. As I discuss shortly, my displacement or dislocation from rootedness in a specifiable landbase has profoundly disrupted my attunement to the more-than-human world. But living in a world abuzz with the fire of life and shot through with deep connections both seen and unseen *are* part of my everyday experience. The problem I face is that I inhabit a culture that continually reinforces doubting, ignoring, and overlooking this. So perhaps the primary danger I face is being unable not just to understand what I experience but actually to *experience* what I experience. I must proceed cautiously.[5]

"Animism is the only world religion that doesn't need to scurry to get aboard the environmental bandwagon," states Daniel Quinn (1994, 167–68). In comparison to the practices of the people of our culture, for animists, "it can be said that *conservation happens* but not because conservation is cognized beforehand and then executed," declare Danny Naveh and Nurit Bird-David (2013, 36). I make this point not to "angelize" animists or to treat indigenous peoples, the main practitioners of animism, as "ecologically noble savages" (Nelson 2008a, 12). Like us, animists have made grievously harmful ecological mistakes (Köhler 2005, 423, and Martinez, Salmón, and Nelson 2008, 90). Unlike us, they are conceptually well equipped to restore their broken agreements with the community of life (Hogan 1995, 11).

Not a Matter of Faith

Animism bears little resemblance to the organized religions with which most of us are familiar. First, there is no such thing as an "animist 'in general'" (Stengers 2012, 9) in the same way that one might identify as a Muslim or a Christian while nevertheless belonging to a specific sect or denomination. Indeed, indigenous peoples have no corresponding word for "animism." English anthropologist Edward Tylor coined the term in 1871 in the process of attempting to categorize testimony about indigenous cultures provided by missionaries and explorers.[6] As Linda Hogan (2013) remarks, what we call animism is, from the perspective of indigenous peoples, more aptly regarded simply as "tradition." So there are as many iterations of animism as there are traditions. Our task is to try to capture key points of intersection among these disparate traditions so that we can construct a loose mosaic out of oft-repeated themes.

Second, while animists do speak of the sacredness of life, this is less what we would conceptualize as a religious expression than a pragmatic statement, notes Rebecca Adamson. "In fact, none of the native languages [of North America] have words or terms synonymous with religion. The closest expression of belief translates to the way you live" (2008, 35). In the same vein, Melissa Nelson states that for animists, "the divine or the sacred is something that's a verb; it is active, it is a cocreative process that Indigenous Peoples and all peoples have a right and obligation to participate in on a moment-to-moment basis through our actions, thoughts, and behaviors" (Martinez, Salmón, and Nelson 2008, 108). The sacred thus is not transcendent. We live, breathe, and, yes, eat it. This is why to be religious in the animistic sense has nothing to do with faith or belief and everything to do with how one engages in even the most mundane interactions with fellow members of the community of life (Forbes 2008, 15, and Mohawk 2008, 134).

A World Full of People

Given that speaking of animism as a religion can be misleading, perhaps we should focus more directly on what the experience of being an animist entails. This will clarify why no one can be an animist in general. The term *animism* has historically been used to refer, often derogatorily, to the belief that spirits inhabit things, and not just living things but also presumptively inanimate things like rocks, rivers, and mountains. Some recent usages of the term tend, unwarrantedly, to convey a human locus. Philippe Descola, for example, takes the core characteristic of animism to be "the attribution by humans to nonhumans of an interiority identical to their own" (2013, 129; see also Viveiros de Castro 1998).

Let us instead acknowledge, states David Abram, that spirits in indigenous cultures are "primarily those modes of intelligence and awareness that do *not* possess a human form" (1996, 13).

Perhaps with this in mind, Graham Harvey defends a considerably more illuminating way that the term *animism* is now used:

> Animists are people who recognize that the world is full of persons, only some of whom are human, and that life is always lived in relationship with others. Animism is lived out in various ways that are all about learning to act respectfully (carefully and constructively) towards and among other persons. Persons are beings, rather than objects, who are animated and social towards others (even if they are not always sociable). Animism *may* involve learning how to recognize who is a person and what is not—because it is not always obvious and not all animists agree that everything that exists is alive or personal [i.e., a person]. However, animism is more accurately understood as being concerned with learning how to be a good person in a respectful relationship with others. (2006, xi)

A person is a being who may be communicated *with*, Harvey asserts, a being with whom other people interact "with varying degrees of reciprocity" (ibid., xvii; see also Sterba 1995, 101). To be a person is to be a self-aware, communicative, volitional, relational, cultural, and social agent. People can act intentionally and autonomously (ibid., 100; see also Hallowell 1960 and Bird-David 1999). They have minds of their own, and they use them. These are all characteristics that both animals and plants exhibit.

"That some persons look like objects is of little more value to an understanding of animism than the notion that some acts, characteristics, and qualia, and so on may appear human-like to some observers" (ibid., xvii). So outward appearance is not a decisive indicator of personhood. Moreover, the characteristics of personhood are not quintessentially human ones that are projected by animists onto other-than-human beings. Rather, we humans share traits and capacities with many other kinds of people. We are "part of a heterarchical continuum of persons," Hall states (2013, 388). Personhood thus is a more general category "beneath which there may be listed subgroups such as 'human persons,' 'rock persons,' 'bear persons,' and others" (Harvey 2006, 18).[7] We are not outliers in this regard. We are not exceptional, nor are the characteristics and abilities that traditionally have been taken as indicators of our distinctiveness. While our closest relationships usually are with other humans, the same surely can be said of other-than-human people (Hallowell 1960). And that our bonds of solidarity with human compatriots are especially intimate does not invalidate our affinity with and responsibility to others.[8]

This is not to say that the identification of others as people is always easy. Indeed, Jean Piaget (1929 and 1954) has it exactly backward when he suggests that animism—denoted according to him by the assumption that every event is the result of a deliberate act—is a standard childhood phase that is outgrown through further and fuller cognitive development. As Harvey notes by way of contrast,

> This animism (minimally understood as the recognition of personhood in a range of human and other-than-human persons) is far from innate and instinctual. It is found more easily among elders who have thought about it than among children who still need to be taught how to do it. In learning to recognize personhood, animists are intended, by those who teach them (by whatever means) to become better, more respectful persons. That is, humans might become increasingly animist (reaching beyond the minimal definition) as throughout life they learn how to act more respectfully (carefully and constructively) towards other persons. (2006, 18)

Note here Harvey's parenthetical reference to the need for careful and constructive engagement with other people. Yes, respect may entail veneration, but this is not always so. Some people want to eat us. Others may prove even more dangerous (Justice 2011, 415). We do well to be cautious of them (Chernela 2001).

Note also that like moral standing, personhood is ours not to confer but instead to discern. This is because, for the animist, humans do not hold some preferred or special ontological status that enables us to attribute personhood even if we want to do so. While individuals within species—often elders—may wield extraordinary power, maintain prestige, or exhibit wisdom, no species corners the market on these capacities. Indeed, keystone species in particular can act as conveyors of knowledge and proper behavior to humans (Buhner 2002, 183).

Note, finally, that it may be the case, as panpsychists suggest, that every particle in the universe is alive and in some sense aware (Plumwood 1993, 133, and Mathews 2003). It instead may be that we must learn to distinguish between people (those communicated with) and objects (those communicated about). It also may be that the distinction between people and objects is permeable and subject to change.[9] The central point is this, states Plumwood: "Monopolizing mind may make us feel superior but it is not helping our accommodation to the earth" (2013, 449).[10] Nor, from an animistic vantage point, does it provide a particularly accurate description of the relationships that exist among members of the community of life.

The Illusory Specter of Anthropomorphism

Are animists ever guilty of projecting human likeness onto other-than-human beings? Surely they are. As Harvey asserts, "it would be foolish to deny that some of what passes for animism might be anthropomorphism." But he stresses that this is why it takes careful attunement to and no shortage of experience with the more-than-human world to avoid "the unwarranted projection of personhood" or "the inaccurate attribution of particular intentions." That "animists *can* mistakenly project life demonstrates that particular versions of animism require education" (2006, 170).

Plumwood acknowledges that the charge of anthropomorphism may perhaps prove useful in some cases when humans fail to respect ways in which other-than-human people are indeed different. Usually, however, this charge "begs the question of non-human minds. That has become its major function now, to bully people out of 'thinking differently.' It is such a highly abused concept, one often used carelessly and uncritically to allow us to avoid the hard work of scrutinizing or revealing our assumptions, that there is a good case for dropping the term completely" (2013, 452).[11] Likewise, Mary Midgley (1983 and 1996) notes that *anthropomorphism* is an ambiguous term. It is sometimes used to attribute to other-than-human beings characteristics that humans have. At other times, it refers to the attribution of characteristics that *only* humans have. Both uses are problematic. The first implies that humans and other-than-human beings share no characteristics worth highlighting. The second dismisses without argument the very point that animists emphasize.

The issue here is not that animists make sense of the living world though human experiences. This "weak" form of anthropomorphism cannot be avoided. Nor is it necessarily harmful. It is the "strong" form of anthropomorphism, the attribution of human meaning onto other-than-human traits and behaviors, that is potentially damaging. It is also avoidable, and errors of attribution are correctable. Surely animists run into "real problems representing other species' communicative powers or subjectivities in terms of human speech," Plumwood asserts, but these problems "do not rule out such representation in any automatic way" (2002, 59). They instead highlight the need for careful attunement to the more-than-human world.

Topocentrism and Relationality

I mentioned earlier that sacred eating is no utopian view. This stems from the animist's inability to be utopian. The term *uto-pia—u*, away from, *topia*, land—connotes disconnection or dislocation. It is that which has never found a place where it is fit to exist. By contrast, animistic traditions are indelibly rooted in specific landbases. Animists are kincentric. They tend to regard their ties with the human and other-than-human people with whom they share their lives and land as sacred bonds with family members, some of whom they must nevertheless engage with cautiously (Quinn 1996, 182f., Weaver 1997, 39, Salmón 2000, and Rose 2013, 137ff.). So instead of focusing strictly on responsibilities to their immediate family or their direct line of descent, they focus on extended expectations to clan and land (Justice 2011, 59). This is essentially the same as saying that they are topocentric. For animists, kincentrism and topocentrism are in fact one and the same concept, one and the same experience, regarded in two slightly different ways. Responsibility for the health and well-being of a place just is responsibility for the health and well-being of the coinhabitants of that place. To be *emplaced* is to be enmeshed in bonds of kinship. "Everything begins and ends with land," Harvey proclaims. "More precisely, particular lands place everything, everyone, and every happening in relation to and in communion with one another." Landbases are "not mere scenery, backdrops, or stages for the grand drama of life; they define, birth, contextualize, and participate. Lands erupt into life and fully engage with the emergent and proliferating diversity" (2006, 66). Moreover, state Catherine and Ronald Berndt,

> No traditional Aboriginal myth was told without reference to the land, or to a specific stretch of country where the incidents it relates were believed to have taken place. No myth is free-floating, without some local identification. [. . .] [T]he land and all within it was irrevocably tied up with the content of a myth or story, just as were (and are) the people themselves. [. . .] It is, then, the land which is really speaking— offering, to those who can understand its language, an explanative discourse about how it came to be as it is now, which beings were responsible for it becoming like that, and who should be responsible for it now. (1989, 5–6)

This is what makes animists more or less default conservationists. The landbase, its health and well-being, is primary. *It* tells *us* which forms of use, of it and of all life on and in it, are acceptable and which are not. *Its* needs must come first if ours are to be met. This is perhaps the core precept of sustainability (Jensen and McBay 2009, 55ff.).

To understand the needs of a landbase—to comprehend its language—requires a deep and abiding familiarity with it. This is no easy task for the people of our culture. As Descola remarks, it remains all too common for our kind to deride, however subtly, "the lazy propensity of barbaric and savage peoples to judge everything according to their own particular norms" (2013, xv). This is due in part to the fact that we are governed by the demands of mobility, of *displacement*. In practical terms, many of us have lost all sense that landbases speak, let alone that we have the capacity and responsibility to listen to them. Additionally, states Plumwood, "both place and the more-than-human sphere are disempowered as major constituents of identity and meaning" for the people of our culture (2000, 231). Regaining this is the task of more than a lifetime, especially given the structural obstacles we face "to developing a place-sensitive society and culture" (ibid., 232).

Plumwood notes as well that we must be especially careful not to let a renewed concern with our landbase lead us to overlook "shadow places" at the expense of places that are beautiful, pristine, or well maintained. Shadow places are local and distant settings that are ecologically despoiled and socially fragmented in large part because they serve as dumping grounds or resource extraction sites for beautiful places. Out of sight, shadow places suffer so that beautiful places can remain beautiful. Precisely because they remain in the shadow of beautiful places, they "elude our knowledge and responsibility," Plumwood asserts (2008, 139). This permits inhabitants of beautiful places to become "more and more out of touch with the *material conditions (including ecological conditions) that support their lives*" (ibid., 142; see also Ehrenreich 2002). They lose all sense of how the labor of other people and of the land directly benefits them.

Any ethics of place thus also must be accompanied by a politics of place, Plumwood argues. Matters of justice—of recognition, redistribution, and restoration—must not be ignored. Yes, seeking self-sufficiency has its benefits. But we must also turn a critical eye to how places are connected and to the power dynamics that these connections harbor (Hayden 1995 and Haraway 1997, 215).

Indeed, it is not merely the case that our lives are indelibly connected to others' lives. For good or ill, *we are these relations.* We "exist as 'corporate elaborations'—composite communities of cells built out of the accomplishments of our one-celled forebears," state Mahlon Hoagland and Bert Dodson (1995, 2).[12] Moreover, we are part of, constituted by, a complex web of interactions among animals and plants. We are utterly dependent for our survival, no less, on the billions of bacteria who inhabit our intestinal tract. We are ecologically embedded animals through and through. This is very important to keep in mind, states Harvey, "as people negotiate daily needs" (2013, 3; see also Naveh and Bird-David 2013, 27).

Our Distant Green Relatives

Matthew Hall (2013) notes that animists have known for millennia what scientists are only now discovering. Phylogenetically, plants and animals are closely related, so we share many developmental attributes (Stiller 2007). It thus should be unsurprising that numerous peoples, including the Maori of New Zealand, the Raramuri of northwestern Mexico, and Australian Aboriginals, tell stories of common descent.

How do we honor our kinship with plants? I mentioned at the opening of Chapter 2 that sentientists mistake our deafness for plants' dumbness. Most of the people of our culture do. But those whom Stephen Buhner refers to as *vegetalistas* do not. The numerous vegetalistas who Buhner interviews in his study of the lost language of plants are "uniformly consistent in saying that their personal and cultural knowledge of the medicinal actions of plants [comes] from 'nonordinary' experiences, specifically: dreams, visions, direct communications from the plants, or sacred beings" (2002, 33). This knowledge is not gained primarily through trial and error. The plants themselves convey to the vegetalistas the ways in which they see to the health of their landbase—how they can serve as medicine for us insofar as they are medicine for all (ibid., 197).

Vegetalistas thus specialize in what we might call *inquisitive* knowledge as opposed to *acquisitive* knowledge. They treat plants—both those who are keystone species and those who are not—as teachers and guides from whom they are privileged to learn, not as resources to be exploited. Cherokee and Creek vegetalistas emphasize to Buhner that the power to provide care lies primarily with plants rather than with humans. In the same vein, a Cahuilla doctor tells Ruby Modesto that interactions with plants must exhibit sincerity and humility, since "Some of them have powerful spirits" (1980, 38; see also Arnold and Gold 2001). This does not preclude the use of plants, of course. This is why negotiation is a critical aspect of vegetalistas' relationship with them.

Consider also the invocations of remembrance and reverence of Maori women when they dig up sweet potatoes, who traveled with the Maori in their canoes when they migrated to New Zealand (Reedy 1997, 121ff.). T. P. Tawhai (1988, 101) contends that these invocations call for neither redemption nor forgiveness for eating the sweet potatoes. Instead, they offer praise and thanksgiving while also seeking permission. (Perhaps this form of engagement with whom we make our food would be more satisfactory to Preece. It seems to circumvent his criticisms, in any case.) And then there are the women of the Yekuana, who live in the Amazon, greeting and singing protective songs to young Yucca plants (Guss 1989, 36). These are but two examples among many, but they indicate quite evocatively just how tightly the bonds between humans and plants can be—tighter even than Plumwood countenances, which is perhaps the primary reason that I favor animism over ecological animalism. Elaborate ceremonies may not be necessary. Some "are almost invisible to outsiders," states Harvey (2006, 166). But they display how important it is to acknowledge plants as both food and more than food.

The Contextual Case for Vegetarianism

My second attempt at a moral defense of vegetarianism, which comes in the form of a care-sensitive ecological contextualism, is informed by three basic propositions that are most easily understood from within an animistic conceptual framework:

1. The landbase is primary.
2. Who we eat is a function of where we eat.
3. How who we eat becomes our food is care-sensitive.

Permit me to elaborate.

The Landbase Is Primary

Activities are ecologically sustainable so long as they either help or do not materially harm landbases. A landbase is a function of the ecosystem of which it is part, and ecosystems are constituted by countless mutually dependent relationships among organisms. Ecosystems thus comprise what Eduardo Kohn (2013) calls an "ecology of selves." They differ from one another and change over time based on the ways in which the selves, the people, who constitute them interact. How these people interact is partially a function of climate and available sunlight. So the particular needs and interests of one landbase are not the same as those of any other, and the needs and interests of each landbase change over time.

If we intend to act in ecologically sustainable ways, we must follow the lead of our landbase. What we do to it, we do to ourselves. This does not entail that we should discount individuals' specifiable needs and interests, but we should acknowledge that they are a function of relations among emplaced coinhabitants. So it makes little sense to address individuals' needs and interests without seeking to understand the contexts in which they arise. And as within any community, colliding agendas among individuals are inevitable. We must be willing to acknowledge and carefully navigate "the messy realities of a shared world," Harvey remarks (2006, 117). This entails accepting that painful trade-offs and difficult decisions are part of life. How these decisions are made may not be entirely up to us, but we must not shirk our responsibility to play our part in making them. This requires that we be aware of which uses of our landbase and of those who inhabit it are sustainable and which are not. So we must develop and maintain an intimate working knowledge of our landbase. We must avoid at all costs imposing our will on it no matter how well intended we may be.

Sustainability is a matter not just of context but also of scale and time. A single person eating in a manner that is unsuitable to his or her landbase is unlikely to cause visible harm. Many people doing so could be catastrophic, especially in the long run (Jensen and McBay 2009, 58).

Sustainability is a function of what other sorts of stressors a landbase experiences as well. A healthy landbase may be able to handle considerable stress. Like us, it can handle much less stress, or perhaps none at all, if its health and well-being are seriously compromised (ibid., 59). Restoration of ecosystems hereby must be among our highest priorities, especially given that landbases the world over face acute stress from global climate change, urbanization, increasingly invasive forms of oil extraction, and exponential human population growth.

Ecological restoration, or what George Monbiot calls *rewilding*,

> is about resisting the urge to control nature and allowing it to find its own way. It involves reintroducing absent plants and animals (and in a few cases culling exotic species which cannot be contained by native wildlife), pulling down the fences, blocking the drainage ditches, but otherwise stepping back. At sea, it means excluding commercial fishing and other forms of exploitation. The ecosystems that result are best described not as wilderness, but as self-willed: governed not by human management but by their own processes. Rewilding has no end point, no view about what a "right" ecosystem or a "right" assemblage of species looks like. It does not strive to produce a heath, a meadow, a rainforest, a kelp garden, or a coral reef. It lets nature decide. (2014, 10)

Monbiot suggests that we start by encouraging retreat from agriculturally marginal and unproductive land, which is already occurring in the United States, Europe, Japan, and elsewhere. Agricultural practices on productive land can be converted from conventional to organic and from monoculture to perennial polyculture, as I will discuss in the next chapter. So we can grow what works for the landbase and avoid adversely affecting soil fertility or overusing available water (Laws 1994). Fragmented ecosystems, a primary driver of species extinction, can be reconnected (Foreman 2004 and Haddad et al. 2015). And desperately needed predators can be reintroduced onto landbases in which they once thrived.

Monbiot places particular emphasis on the last of these phenomena. This is because predator reintroduction is vital for the restoration of trophic diversity, which "enhances the number of opportunities for animals, plants, and other creatures to feed on each other, to rebuild the broken strands of the web of life" (Monbiot 2014, 84). One prominent example of this is the reintroduction of wolves into Yellowstone National Park in 1995. This has been pivotal for restoring an entire ecosystem, Monbiot argues, largely because the wolves prevent overgrazing by the large elk population (Bechta and Ripple 2006 and Ripple and Bechta 2012). Collaborations between pastoralists and predators—as occurs today in southern Africa, Mexico, Chile, and the Upper Midwest of the United States—also can reverse desertification by preventing overresting, or the formation of a hard crust on the soil that keeps out air and water and prohibits seed germination (Savory 2013). Together pastoralists and predators can keep herds tightly bunched and regularly on the move, which plays an irreplaceable role in breaking up and aerating the soil. The herds also provide a judicious flow of nutrients for the soil through dung deposits (Zimov et al. 1995, Pollan 2006, 323, Schwartz 2013, and Monbiot 2014, 90ff.).

Who We Eat Is a Function of Where We Eat

Given the primacy of the landbase, we should obtain our food in ways that foster its health and well-being. This is the same as saying that we should obtain food in ways that are responsible to the health and well-being of the coinhabitants of our landbase so far as colliding agendas permit. Within some contexts, "letting nature decide" may well go hand in hand with many or most people embracing vegetarianism, or at least greatly reducing the use of animals as food. Within others, it may be incompatible with all but a very small number of people being vegetarians without engaging in clear-cutting or slash and burn—neither of which is ecologically viable in the long run (Benhim 2006 and Savory 2013, 75). And some locations simply are not fit to feed us at anywhere near the population levels we may request. Perhaps *no* location is at present, given how many of us humans there now are. But some places, like the Sonoran Desert in the American Southwest, stand out in this regard. "The Hohokam and Pima were the last people to live on that land without creating an environmental overdraft," notes Barbara Kingsolver (2007, 5).

These dictates do not rule out taking account of circumstances in which individuals may fare better as vegetarians or omnivores even if a large percentage of the human population eating as such would materially harm the landbase. This applies, for example, to infants, children, gestating and lactating women, the elderly, and the ill (Warren 2000, 130). Since who we eat coming from where we eat can provide us with a more direct experience of the costs of feeding ourselves, we also can negotiate and compromise with coinhabitants to better facilitate accounting for individuals' idiosyncratic needs. This enhances our ability to refrain from causing the landbase harm. It also can help prevent food waste, which is nothing other than the waste of life.

Seeing to it that who we eat comes from where we eat also permits us see more clearly where our responsibilities to the community of life lie. Jensen and McBay provide a particularly helpful way of expressing this: "if I consume the flesh of another (or otherwise kill this other) I now take responsibility for the continuation of the other's community. This deal holds morally, it holds spiritually, and it certainly holds physically. Those who do not know this—those who do not live it—do not survive. They destroy their own habitat, and in doing so, destroy themselves. It may take a while for those circles to close, to become self-made nooses around their own necks, but it happens. Every time" (Jensen and McBay 2009, 55; see also Jensen 2006, 179).[13] We accept responsibility to do well by those who sustain us. We likewise accede to the fact that our survival depends on the messy realities of a shared world. Respect for life cannot come without respect for death and how it is carried out (Kover 2010, 75). Once again, it can be no other way.

According to Plumwood, the respectful treatment of animals necessitates a great reduction in "first-world meat eating" (2012, 78). Moreover, there are "plenty of good reasons for being a vegetarian in most urban contexts" (2000, 289). But, she asserts, because it is all but impossible in most locales to be a vegan without relying on the global marketplace, its practitioners tend to defy ecological accountability (2004, 354f.; see also Pollan 2006, 326).

How Plumwood reaches these conclusions does not matter for our purposes. I must admit that I do not follow her reasoning anyway. What is important for us is that her claim about veganism may be incorrect for reasons that are closely tied to what makes—or, more accurately, has the potential to make—urban contexts particularly suitable for vegetarianism.

Cities are places that more and more of us are calling home. Demographers estimate that by 2050 more than 70 percent of world's human population will live in urban settings (Cohen 2003 and Buhaug and Urdal 2013). And cities currently demand ever more not just from their own landbases but from *all* landbases. So giving some thought to how to reduce their ecological footprint is of utmost importance.

As is now quite popular in my home city of Philadelphia, residents can convert vacant lots—of which there is no shortage—into community gardens and urban farms (Ladner 2011 and Cockrall-King 2012). Along with providing ready access to fresh produce, community gardens and urban farms provide a good avenue for residents to develop a more engaged, knowledgeable, and caring relationship with their landbase. But vertical growing in high-rise farms provides a sounder path for scaling up urban food production.[14] While vertical farms can raise fish, crustaceans, and mollusks (while using their waste for fertilizer), they are most efficiently used to grow plants. This suggests that the widespread embrace of vegetarianism, and perhaps even veganism, by city dwellers may well facilitate urban sustainability, at least with respect to food production and consumption.

Vertical growing is soilless. Longtime advocate Dickson Despommier highlights that it is carried out via drip irrigation, hydroponics, and aeroponics (2010, 162ff.).[15] It makes efficient use of limited space to produce fresh food locally, which almost entirely eliminates fuel use associated with transporting, storing, and packaging produce (Ladner 2011, 67, and Grewal and Grewal 2012). It uses far less water than does conventional farming. Runoff is eliminated, and herbicides and pesticides are unnecessary.

Growing produce in vertical farms can take place year round without losses from natural disasters, disease, infestation, or spoilage. This is a major advantage, especially given the inevitability of increasingly extreme and erratic weather due to global climate change. To meet their energy needs, vertical growers have the potential to reprocess human waste, using gray water for irrigation and incinerating solid remains on site to produce electricity (Despommier 2009). While inflated property prices may make locating centrally in cities difficult, vertical growing can be implemented seamlessly in cities' shadow places. This has occurred already with the construction of The Place in Chicago (Cockrall-King 2012, 265), which also provides employment for residents and has reinvigorate a sense of the importance of a part of the city that has long been neglected (Despommier 2010, 25).

Vertical growing has another significant advantage. To the extent that it can supplant cities' reliance on land-based agriculture, Despommier asserts that it can facilitate the rewilding of farmland, which "is the easiest and most direct way to slow climate change. These landscapes naturally absorb carbon dioxide, the most abundant greenhouse gas, from the ambient air. Leave the land alone and allow it to heal our planet" (Despommier 2009, 80; see also Despommier 2010, 135ff.). Such developments have the potential to mitigate the effects of global climate change (Bonan 2008). They also can reverse the destruction of landbases the world over. In urban contexts, then, vegetarianism and perhaps even veganism may be compatible not just with urban sustainability but also with widespread ecosystem restoration.

Although vertical growing is being implemented, among other locales, in Singapore, Japan, the Netherlands, and the United States, it is still too new to know whether it can prove successful the world over.[16] Nor will its ecological effects be fully understood for years to come (Despommier 2013, 389).

But it shows great promise, and it is certainly compatible with a considerable reduction in first-world meat eating.

How Who We Eat Becomes Our Food Is Care-Sensitive

Bernice Fischer and Joan Tronto contend that care is "*a species activity that includes everything we do to maintain, continue, and repair our 'world' so that we can live in it as well as possible*. That world includes our bodies, our selves, and our environment, all of which we seek to interweave in a complex, life-sustaining web" (1990, 40). This conceptualization of care fits nicely both with regarding our landbase as primary and with who we eat being a function of where we eat. It also suggests that our relationship with who we make our food should entail one or more of the following: *caring about* them in the sense of taking responsibility for the continuation of their community, *caring for* them individually, *being careful* of them, and being willing to *receive care* from them. Having a care-sensitive relationship with who we make our food thus bears a strong resemblance to sacred eating. It involves being attuned to, responsible for, and responsive to others when one is involved in what Plumwood calls an "exchange of life."

This is not simply a romantic notion. It is eminently practical. Plumwood proclaims that there is no context in which factory farming, or the maintenance of "flesh factories" (2000, 289), is suitable. Shutting them down would foster a great reduction in first-world meat eating, in large part by making eating animals more expensive. So "maybe when we did eat animals we'd eat them with the consciousness, ceremony, and respect they deserve," Pollan suggests (2006, 333). Shutting down factory farms is also compatible with ecological restoration, since so much of the plant life grown via conventional agriculture is used as feed for factory-farmed animals.

As I began to explore in Chapter 2, there are also excellent reasons associated with the health and well-being of plants to bring an end to conventional agriculture. Indeed, there are no contexts suitable for *flesh fields*, or the vast tracts of scarred, denuded, and poorly treated land that are the byproducts of conventional agriculture. Consider, for example, that conventional agriculture exposes plants raised to be crops to large quantities of antibiotics not just through effluent flows from runoff but also through direct treatment (Martinez 2009). Just as animals in flesh factories are susceptible to infectious diseases due to their atrocious living conditions, plants in monocropped fields are at much greater risk of infection than they are on "unkempt" landbases. Direct application of antibiotics "kills not only bacteria on the plants but all susceptible bacteria in the soil itself, with cascading effects on soil integrity and health," Buhner remarks (2002, 130). This has the potential to damage not just single fields but entire ecosystems when spread by wind and water.

So the use of antibiotics makes the already poor living conditions for plants in flesh fields even worse. Yes, they have been genetically manipulated to tolerate living in homogenous communities. But they nevertheless maintain strong connections with other inhabitants of their landbase. Notable are their ties to plant pollinators. Along with the widespread use of neonicotinoid pesticides, antibiotic use may be partially to blame for the mass die off of both wild and domestic honeybees in North America, Buhner contends. This leaves monocropped plants increasingly socially isolated, and it compromises their ability to reproduce—if they aren't grown from sterile seeds.

Consider as well how flesh fields disrupt the mutually beneficial relationship between plants and soil bacteria. Just as we rely on the bacteria in our intestines to metabolize food and bacteria rely on the food we ingest, plants rely on bacteria to facilitate the decaying process that releases the minerals and organic substances they need to grow. Plants not only shed roots and leaves that nourish bacteria; they also excrete complementary minerals and organic substances through their roots (Neal et al. 2012, and Neal and Ton 2013). "In essence, they are soliciting companionship from specific kinds of bacteria," Jensen and McBay declare. They highlight that corn plants grown in sterile environments excrete far more sugar than is typical. "We both found that profoundly sad. The corn was not in its community—in common language it was lonely—and by excreting all that sugar it was calling out for bacterial companionship. Because it was in a sterile environment, it was calling out in vain" (2009, 13). Monocropped corn is not in an entirely sterile environment, of course. But considering what antibiotic use does to the soil, the experiential difference may be comparable.

I wonder, too, whether vertical farming is entirely compatible with exhibiting care for plants. It is hard to imagine that plants are able to enjoy a good life if they are denied sunlight and soil, the two basic sources of energy that they have evolved to crave. Moreover, Despommier highlights that plants can be grown more densely when cultivated vertically than when cultivated conventionally. He also lauds the capacity to be "in total control of conditions necessary to achieve optimal survival, growth, and maturation of any given crop, thereby ensuring maximum yield per square foot of growing space" (ibid., 388). This runs up against calls by animists to consider ourselves not managers or stewards for the community of life but instead its supplicants. And it permits plants to be treated like mass-produced commodities rather than as people. From such a perspective, plants essentially become sacrificial lambs for the sake of ecological restoration, which is neither necessary nor preferable—for them or for us.

But nothing in principle bars the creation of healthy and enjoyable homes for plants. Indeed, it is quite possible to integrate human and plant communities, which can enhance our ability to see our food as more than food and improve the quality of plants' lives. Despommier himself acknowledges that "The best way to look at the future of vertical farming is to integrate the production into the buildings in which you live." This is taking place already in Israel, he notes. "They have new buildings in Tel Aviv that have growing systems off the balconies of each of the floors" (Zerbisias 2011).

So a care-sensitive ecological contextualist defense of vegetarianism appears viable. The practice of vegetarianism on landbases suited to it can greatly benefit the community of life. Moreover, eradicating both flesh factories and flesh farms is a necessary condition for respecting the equal moral standing—indeed, the full-fledged personhood—of our food. The aspects of expansionary sentientism that also exhibit respect for the community of life certainly need not be abandoned. But we are not saddled with vestiges of human mastery either. This is promising.

CHAPTER 4

THE CLOSED LOOP

But I wonder, is a contextualist defense of vegetarianism actually a defense of vegetarianism at all? As I specified in Chapter 3, we must take the landbase as primary. We should listen to it when it tells us which uses, of it and the many kinds of people who inhabit it, are acceptable and which are not. This is our top priority, and it is perhaps the basic precept of ecological sustainability. If abstaining from eating animals is amenable to the needs and interests of our landbase, then we have reason to embrace it. We can articulate this ecologically based decision in moral terms if we wish (Keller and Golley 2000). But we can just as easily regard it pragmatically, for calling it *right* amounts to little more than praising a way of eating that works within a given context. And if it does not work, then we must not engage in it—whether or not we call it *wrong*. This indicates that nothing about vegetarianism per se is morally relevant. Whether we should endorse it is entirely a function of the needs and interests of our landbase.

Furthermore, I now wonder whether *any* defense whatsoever of vegetarianism is possible. In other words, I am no longer certain that one even can *be* a vegetarian. Eating is a transitive property, or so I contend. A transitive property is such that if one element in a sequence relates in a certain way to a second element and the second element relates in the same way to a third, then the first and third elements relate in the same way as well. As an updated version of the old saying goes, we are who we eat. Our food is who our food eats, too. As a result, we are who our food eats in equal measure. As members of the community of life, we and our food are both part of a closed-loop system. Plants acquire their nutrients from the soil, which is composed, among other things, of decayed plant and animal remains. Yes, plants feed on animals. So even those of us who might otherwise believe that we subsist solely on a plant-based diet actually eat animal remains as well. This is what makes being a vegetarian impossible. Arriving at this conclusion should be neither particularly surprising nor all that troubling when regarded from an animistic perspective, since animists value having a care-sensitive and land-sensitive relationship with whomever we make our food.

Does this imply that vegetarians are actually omnivores then? No, it does not. From within a closed-loop system, the very distinction between vegetarianism and omnivorism breaks down. Of course, we can distinguish between the last strand in the food web leading to our mouths and all strands that came before. We can classify our alimentary identity according to whether the strand that leads directly to us is a plant or an animal. But this strikes me as ontologically arbitrary and kingdomist. It reflects a denial of the transitive relations that compose the community of life. The animal or plant who we eat is itself constituted through the landbase by the flesh of both animals and plants—as well as fungi, insects, prokaryotes, and a staggering array of the earth's minerals. We are all so constituted, every single one of us. Once again, it can be no other way.

In what proceeds, I first lay out more fully what makes the community of life of which we are all part a closed-loop system. Next, I expand on what I mean when I claim that we are who we eat. While the phrase from which it is derived—"you are what you eat"—is generally intended to indicate that our health and well-being are closely tied to our dietary choices, its historical origins are considerably richer than this. So is what I mean by the phrase "we are who we eat." Indeed, to be who we eat has interlocking biochemical and spiritual aspects, both of which are integral to the transitivity of eating.

I then argue that the transitivity of eating entails that the case for vegetarianism per se vanishes. So long as we are who we eat, we cannot be vegetarians. Nor, for that matter, can we be omnivores. Distinguishing between vegetarianism and omnivorism serves no salient purpose. It is not that this distinction in and of itself is harmful for the planet. Rather, it is emblematic of a way of thinking about ourselves and our place in the world that is deeply rooted in our ecocidal culture.

Finally, I evaluate considerations offered by Michael Fox, who offers support for the proposition that omnivorism and vegetarianism nevertheless may be semantically intransitive. What eating meat and eating a plant-based diet have come to signify contrast starkly, Fox avers. Distinguishing between vegetarianism and omnivorism remains important, according to him, since it permits us to see how embracing vegetarianism can help us to combat oppressive practices that are deeply connected with meat eating. I am in complete agreement with Fox that the oppressive practices he highlights must be combatted. But I suggest that they do not reflect something malignant about or intrinsically problematic with meat eating. Indeed, the defense of semantic intransitivity is a manifestation of the sentientist's benighted understanding of how we relate to our food.

THE CLOSED LOOP AND THE COMMUNITY OF LIFE

Each and every living being on Earth belongs to the community of life. We are bound together in a complex web of relationships marked by interdependence among individuals, species, and entire ecosystems. So we all have roles to play, however minor, in sustaining the health and well-being of this community. We shape it, and it shapes us. We are who we are because of it. And the process of shaping and being shaped is fueled in turn by each member of the community of life feeding and being fed upon. "Nothing is exempt. Nothing is special. Nothing is wasted. Everything that lives is food for another," Quinn declares (2001a, 23f.). Yes, there are better and worse—much worse—ways to become food for others. The obligation to kill well is part and parcel with the obligation to eat well (Haraway 2007, 296). This is but one reason to reject conventional agriculture in all its forms, including both flesh factories and flesh fields. It also should give us a better sense of why maintaining care-sensitive relationships with who we make our food is so vital.

Among the most important of the roles we all play is to pay back upon death the life we have been given by our community. Every life is "on loan" in this respect, and every loan is paid back in full (Quinn 1996, 163). Quite literally, one way or another, we return to the land, states Aldo Leopold: "Land, then, is not merely soil; it is a fountain of energy flowing through a circuit of soils, plants, and animals. Food chains are the living channels which conduct energy upward; death and decay return it to the soil. The circuit is not closed; some energy is dissipated in decay, some is added by absorption from the air, some is stored in soils, peats, and long-lived forests; it is a sustaining circuit, like a slowly augmented revolving fund of life" (1949, 216; see also Callicott 1989). Solar energy flows throughout all biota, and heat is absorbed or dissipated throughout all life processes. The molecules and nutrients that are our building blocks and that become enlivened by sunlight and heat are cycled and recycled. This *nutrient cycle* is what makes the community of life a closed-loop system (McDonough and Braungart 2002, 92). All outputs are used also as inputs. Waste for one is food for another (Commoner 1971 and Hawken 2010, 47). Borrow nutrients, use nutrients, and return nutrients, over and over again.

Indeed, state Hoagland and Dodson, life is characterized by the circular flow of both information and materials. The ongoing process of making and breaking the molecules that serve as nutrients is "the basic process of life" (1995, 144). It is a function of self-regulating and self-correcting feedback. "Life loves loops. Most biological processes, even those with very complicated pathways, wind up back where they started" (ibid., 21). The circulation of blood, the sense-and-response process of the nervous system, menstruation, migration, and the cycle of birth and death all operate on the same principle. So do our bodies for that matter, since every day roughly 7 percent of our cells "turn over," as Hoagland and Dodson put it. Each of us decomposes and recomposes in our entirety every two weeks— although thankfully not all at once.

Ecosystems operate in exactly the same way. The constituent parts of each ecosystem are so tightly interconnected that looping patterns in one affect those in others. Nothing lives, and nothing occurs, in isolation. Together, each smaller loop is thus part of bigger loops. We learn in primary school that oxygen, which is necessary for animals' respiration, is given off as a waste product of photosynthesis by plants. Animals exhale carbon dioxide that is inhaled by plants to make sugar, which they use for food. This process has global climactic effects. Likewise, I recall learning during my secondary schooling that in freshwater systems, fish eat algae and excrete organic waste. Bacteria then eat the waste and excrete inorganic material, which is eaten by algae. If one link in this chain is broken, the health of the entire water system is adversely affected. So is the ecosystem in which the water system is located. And while the people of our culture often struggle to accept that we humans have integral ecological roles to play, our ancient ancestors would not have spread so expeditiously to nearly every corner of the earth many millennia before the dawn of civilization had they not been able to give back in consonance with what they received from so many different kinds of landbases. "From the standpoint of the whole ecosystem, these interchanges occur so smoothly that the distinction between production and consumption, and between waste and nutrient, disappears," state Hoagland and Dodson (ibid., 22). Such interchanges often are noticeable only when some sort of disruption occurs.

The People of the Deer

Like the people of many Native American nations, the Ihalmiut, who once thrived in Canada's northernmost climes, are now all but extinct. Dependent entirely on caribou for their

survival, their numbers plummeted when the caribou population went into steep decline with the arrival of the people of our
culture midway through the twentieth century. The Ihalmiut
were known as "the People of the Deer" (Mowat 1952). Those
who survive still are, at least symbolically and in terms of their
means of self-identification.

To illustrate in greater depth what it means to be a member
of the community of life, Quinn assumes the Ihalmiut perspective, offering a rendering of what they might say about their
connection to the caribou, or deer:

> The fire of life that once burned in the deer now burns in us, and we
> live their lives and walk in their tracks across the hand of god. This is
> why we're the People of the Deer. The deer aren't our prey or our
> possessions—they're us. They're us at one point in the cycle of life and
> we're them at another point in the cycle. The deer are twice your par-
> ents, for your mother and father are deer, and the deer that gave you its
> life today was mother and father to you as well, since you wouldn't be
> here if it weren't for that deer. (1996, 182; see also Quinn 2001b, 86)

Our paths and the paths of our food converge. All our paths are
strands in the food web—the web of life—which Quinn char-
acterizes as the tracks crisscrossing the "hand of god." None is
disconnected. None is self-sufficient. There are no exceptions.
And all constitute the life of the landbase, of a particular place.

Each of us thus is ablaze with the fire, the energy, of life initially provided by the sun. "To each of us is given its moment
in the blaze," Quinn proclaims, "its spark to be surrendered to
another when it's sent, so that the blaze may go on. None may
deny its spark to the general blaze and live forever—not any at
all. [. . .] Each—each!—is sent to another someday" (1996,
187–88). The grass to the grasshopper, the grasshopper to the
sparrow, the sparrow to the fox, the fox to the vulture, and the
vulture (along with remnants of them all) to the grass: all have
their moment to borrow and to return.

In this respect, the Ihalmiut could just as easily say that their
mother and father are grass. For the grass and the deer are "one
thing" whose "name is fire," Quinn contends (ibid., 187). This
is borne out as well, Martin Pretchel asserts, among the Maya

through their concept of *kas-limaal,* or "mutual indebtedness, mutual insparkedness." To recognize and accept *kas-limaal* "is an adult knowledge," a Mayan elder tells Pretchel. To want to be free of debt for the fire of life of which we are part "means that you don't want to be part of life, and you don't want to grow into an adult" (2004, 349). This is deeply disrespectful not just to the community of life but also to all who have died— including our food *and* our ancestors—so that we can live.

Venerating the Dead

What if we were to see death, states Plumwood, "as recycling, a flowing on into an ecological and ancestral community of origin?" This provides a fitting extension in death of what the landbase is in life: "a nourishing terrain." And it permits us to see death as "a nurturing, material continuity/reunion with ecological others" (2012, 92). For each individual, death confirms our transience. But as members of the community of life, it affirms that we are part of "an enduring, resilient cycle" (ibid., 95).

Among many of the people of our culture, our tendency to deny that we ourselves are—must be—food for others is most strongly exhibited by our burial practices. Upon death, our bodies are sealed in a coffin and sarcophagus well below the reach of most roots (Plumwood 2004, 348). The embalming process, originally carried out with arsenic and now with formaldehyde, turns us into deliverers of biocides that kill the microbes that would otherwise facilitate our decay. Under these conditions, the body "can't compost aerobically," state Jensen and McBay. "Instead it putrefies, gradually changing into a semi-liquid residue doped with toxic preserving agents" (2009, 142).[1] This may not be a matter of much concern for those who presume that death entails the departure of our essential selves from Earth. From this viewpoint, death represents separation rather than return. Once we are gone, we are gone. There is no need to care about what happens to our bodies, let alone the world we depart.

But what if once we are gone, we stay? What if whatever makes us *us* returns to the landbase in death just as we are interconnected with it in life (Gunn Allen 1979)? To accept this proposition would alter quite profoundly the way we understand not just our legacy and how we should conduct ourselves while alive but also our connection to those who have come before us. Consider how Abram describes traditional forms of ancestor veneration: "For almost all oral cultures, the enveloping and sensuous earth remains the dwelling place of both the living *and* the dead. The 'body'—whether human or otherwise—is not yet a mechanical object in such cultures, but is a magical entity, the mind's own sensuous aspect, and at death the body's decomposition into soil, worms, and dust can only signify the gradual reintegration of one's ancestors and elders into the living landscape, from which all, too, are born" (1996, 15). The earth is a dwelling place for both the living and the dead, Abram emphasizes. The practice of venerating our ancestors is thus intimately tied to our veneration of other-than-human people, including both plant and animal people who may become our food. It signifies, continues Abram, "not so much an awe or reverence for human powers, but rather a reverence for those forms that awareness takes when it is *not* in human form, when the familiar human embodiment dies and decays to become part of the encompassing cosmos" (ibid., 16). It is perhaps incorrect then to say that denying *kas-limaal* is disrespectful to our food *and* our ancestors, as if these entities are so easily separable. Acceptance of the "adult knowledge" that *kas-limaal* constitutes perhaps should lead us to see that veneration for our ancestors *just is* veneration for our food, for the life-giving processes that bind together the living and the dead both physically and spiritually.

WE ARE WHO WE EAT

There are at least six different ways that the phrase "you are what you eat" has been conceptualized. (1) Its original roots may trace back to early Christianity (Gilman 2008), whereby it makes reference to the conversion of bread and wine into the body and blood of Christ through the Eucharist. On this view, one may commune with Christ—by ingesting him—in the process of memorializing his sacrifice on the cross.[2]

(2) "Tell me what kind of food you eat," declares Jean Anthelme Brillat-Savarin, a pioneer in the field of gastronomy, "and I will tell you what kind of man you are" (2011, 15). What one eats is emblematic of one's station in life. Moreover, we can understand—and perhaps even transform—people's character and dispositions by analyzing and manipulating their eating habits. For Brillat-Savarin, states Isil Celimli-Inaltong, "Food is much more than simply sustenance, it is the lens through which one can understand society and detect social dynamics" (2014, 1847). Indeed, at its best, gastronomy can provide insight into how to cultivate a more informed and sophisticated citizenry, hence to facilitate fulfillment of the basic ideals of the Enlightenment.

(3) For Ludwig Feuerbach, what one eats provides a valuable means to understand the world and one's place in it. Celimli-Inaltong notes, in the process of elaborating Feuerbach's position, that "not only does eating connect people to their surrounding world in ways in which other experiences cannot, but they simply have to eat not just to survive but also to be able to contemplate. The emphasis on food allowed the philosopher to unearth the indissoluble link between the mind and the body, the spirit and nature" (ibid.). To eat well thus has epistemological significance. Proper eating makes for clear thinking. Given that human needs and interests are primary, according to Feuerbach, clear thinking makes possible "deploying nature" to secure one's livelihood. This in turn "allows philosophy to locate its true task by placing the human being at the epicenter of this enterprise" (ibid., 1848). Human mastery, via the internalization of the external world through consumption, is our highest calling (Turner 1996, 185). Eating is the most immediate form of consumption. And eating well permits us to make the world ours—literally *to make it us* by reshaping it in our image.[3]

(4) As I mentioned earlier, the phrase "you are what you eat" is most readily identified today with the idea that the food

we ingest has a powerful effect on our health and mood. So the general connotation is that healthful eating facilitates well-being. Again citing Celimli-Inaltong, "The phrase also refers to the ways in which the chemical structure of the foods people consume affect the way in which they act. By establishing a link between psychology and chemistry, this perspective attempts to explain the association between food intake and human behavior, such as the relationship between impulsive behavior and sugar intake" (ibid., 1845). This suggests that what we eat is a direct reflection of our character as consumers as well. So deploying the phrase "you are what you eat" provides means to blame people for poor eating habits and ill health that may be associated with these habits.

For his part, Fox asserts that "you are what you eat" has a double meaning. (5) He cites Karl Marx, who states that "nature is the *inorganic body* of man" (Fox 1999, 24; see also Marx 1997, 293). Fox takes this rather obtuse claim to mean that nature is us and we are nature, since the inorganic for Marx is that which is not currently a manifestation of our organic body but which can become part of it. (6) Fox also cites Ed McGaa, or Eagle Man, of the Oglala Sioux, who proclaims that "Every particle of our bodies comes from the good things Mother Earth has put forth. Mother Earth is our real mother, because every bit of us truly comes from her, and daily she takes care of us" (ibid.; see also McGaa 1990, 203). So you are what you eat *physically* because of what you ingest and digest. You also are what you eat *figuratively* because your food has symbolic meaning for you and is integral to the construction of your identity and sense of place.

I have no interest in allying myself with early Christians, Brillat-Savarin, Feuerbach, or proponents of what can perhaps be called ethical consumerism. And we can debate the finer points of Fox's specification of the phrase "you are what you eat." But he nevertheless is adamant that "No one can deny the truth of this insight" (ibid.). I do deny it, though. For as it stands, the phrase fails to capture with sufficient depth and richness what it means to be intertwined in the food web with people who we eat and people who eat us. *You* are not *what* you eat. *We* are *who* we eat. I create conceptual barriers between myself and who I make my food if I focus only on you and your relationship with your food. I am no passive observer of food-ways. I am enmeshed in them. And it should be abundantly clear by now why my food is not a what—a mere object or thing.

Postmaterialism

To explain how I conceptualize the phrase "we are who we eat," I begin by drawing on considerations offered by Freya Mathews that initially may seem tangential. Bear with me.

Mathews contrasts three basic modalities that, on her account, inform how the people of our culture understand and treat our world.[4] The modality according to which we live "colors everything we do—our entire culture takes its cue from it," she asserts (2006, 85). The first modality, premate-rialism, operates in accordance with the basic principle that all we are and all we do is dependent on a supernatural source of authority. It is most clearly on display, Mathews asserts, in the monarchies of medieval Christendom, the political system based on the caste tradition in India, and some contemporary regimes in the Middle East. Proponents of the second modal-ity, materialism or modernism, valorize instrumental reason. For materialists, Mathews argues, the most effective use of instrumental reason is to gain "increasing technical control of the environment" (ibid., 91). This is a moral calling for them, and it correlates with their mechanistic view of the uni-verse. The cosmos is indifferent to human concerns and has no concerns of its own. "Humanity has therefore to invent its own reasons for living—its own meanings and values. In the absence of religious revelation, human nature itself becomes the sole source of meaning and value: *humanism* replaces religious value systems and human self-reliance replaces the importunate attitude" (ibid., 90).

Mathews insists, however, that the cosmos is not norma-tively neutral for postmaterialists. But, unlike prematerialists, postmaterialists hold that the basis of normativity does not derive from a transcendent source of authority. The normativity of the cosmos derives "rather from an inner dimension of matter itself. In other words, postmaterialism does not posit the supernatural—in the sense of a realm that lies beyond nature—but discovers normativity within material reality itself. This normativity emanates from a dimension of material reality that is in principle unobservable, and hence cannot be revealed by science. It is in principle unobservable in the sense that it consists of the interior aspect of matter" (ibid., 93). The material world bears "a kind of subjectivity or mentality or inspiritment or conativity" (ibid., 93–94). For this reason, the universe itself is sacred, and its sacredness derives entirely "from matter's own inner principle" (ibid., 94).

This view of the cosmos is entirely consistent with the findings of science, but—significantly—Mathews insists that it is not exhausted by these findings (see also Kaufman 2008 and Nagel 2012). In consonance with animism, matter's "inner principle" is never revealed once and for all. It cannot be made legible for inscription in sacred texts to be used by mundane authorities for social control. Nor can it be harnessed as a means to gain technical control, or what is really the *illusion* of technical control, over the world. Rather, Mathews continues, "we must simply try to accommodate ourselves to its 'givenness' in any particular situation. This means trying in every situation to detect the contours of its unfolding and accommodating our agency to its contours" (ibid.); "the normativity of postmaterialism has to be decoded afresh in every situation and cannot be anticipated by rules or mediated by authorities" (ibid., 95). This is why we cannot—and must not try to—impose our will on the land but instead must attune ourselves to its needs and interests.[5]

Biochemical and Spiritual Transference

What bearing does postmaterialism have on the proposition that we are who we eat? Being who we eat has a biochemical basis, which has been subject to extensive scientific investigation. It also has a spiritual basis, interlocked with and in some respects indistinguishable from the biochemical basis, which has been given much less attention by the people of our culture. So being who we eat involves both biochemical and spiritual forms of transference from those whom we make our food to us.

With respect to biochemical transference, biologists have provided clear evidence of the influences of food on the composition of body tissue (Kohn 1999). Corey Bradshaw and his colleagues (2003) have identified seasonal and regional differences in the adipose tissue of southern elephant seals, which has permitted them to map the elephant seals' foraging behavior. Research by Dangsheng Liang and Jules Silverman (2000) indicates that nestmate recognition among social insects is governed in part by chemical cues. Among Argentine ants, hydrocarbons function as these cues, and the molecular structure of these hydrocarbons is the same as that which is present in the ants' diet. So who Argentine ants eat plays a pivotal role in the function of their social lives.

We humans are no different. The kinds of bacteria that populate our intestines differ in accordance with our long-term eating habits (Wu et al. 2011). Moreover, in comparison to our ancestors, the bodies of many North Americans today contain an outsized quantity of the carbon 13 isotope, which comes from the ubiquity of corn in our diets. "When you look at the isotope ratio," remarks Todd Dawson, "we North Americans look like corn chips with legs" (quoted in Pollan 2006, 23). Just about every product sold in any given convenience store contains high-fructose corn syrup. Feed corn and corn silage makes up a considerable amount of the diet of factory-farmed cattle, pigs, and chickens. So whereas the Ihalmiut are the People of the Deer, it would seem that we are the People of the (Genetically Modified) Corn. But there is a big difference between the Ihalmiut and us, of course. Most of us are utterly blind to our constitution. The Ihalmiut are not blind to theirs.

We are not who we eat through and through, of course. Each of us is genetically unique. But even this is complicated by epigenetics. Our nutrition influences whether our offspring, and our offspring's offspring, are genetically predisposed to suffer from a wide variety of ailments (Susiarjo and Bartolomei 2014). We can even influence our own genetic expression—we can "switch genes on and off," as Jacqueline Vanacek (2012) puts it—through our dietary choices. So the way our genes behave corroborates that we are substantively who we eat.

What of the spiritual transference associated with being who we eat? Animists identify at least two forms that it takes. I mentioned in passing that ancestor veneration is inseparable from veneration for our life-sustaining food. Nourishment and nurturance of our lineal relationships thus go hand in hand. I say more about this shortly.

Spiritual transference also occurs more directly, according to people like the Runa of the Ecuadorian rain forest and the Nama of South Africa. The Runa maintain deep kinship ties with the jaguar. They report that it is not uncommon for humans to be transformed into jaguars with white hides. Specifically, they become shape-shifting human-jaguars, or *runa puma*. "Men drink jaguar bile to become puma," Eduardo Kohn remarks. Their consumption of the bile permits the men "to acquire some of their [i.e., jaguars'] selfhood" (2013, 119). For the Nama, it is not necessarily some of the selfhood of their food that they acquire but their food's quintessential abilities or traits. For instance, the Nama refuse to eat rabbits because they hold that this makes them fainthearted. But they do not hesitate to drink lion blood, because it makes them strong and agile. "A vast number of human cultural traditions point us to this idea," Ragan Sutterfield asserts, "that to eat something is to meld our being with it and it with us" (2012).

We can dismiss ideas like these as archaic superstitions, of course. Or we, too, can go beyond thinking of our food in terms of caloric intake, carbohydrate content, or what have you. We can identify our pursuit of health and well-being with the facilitation of a food web that is itself vital and vibrant. We can become conscientious students of the lived experiences of who we make our food, whether or not we envision being able to acquire some of their selfhood or take on their abilities or traits.

But I would not be so quick to dismiss that the transference of lived experiences via ingestion and digestion is indeed possible. Recall that we are our relations, according to animists. I mentioned in Chapter 3 that, according to Hoagland and Dodson, you and I "exist as 'corporate elaborations'—composite communities of cells built out of the accomplishments of our one-celled forebears" (1995, 2). Nurit Bird-David emphasizes that "We do not personify other entities and then socialize with them but personify them *as, when,* and *because* we socialize with them" (1999, 77). And, Harvey insists, "Subjectivity itself is communal and continuously expressed in action" (2006, 113). Given the central role it plays in each of our lives, one means of expressing our communal subjectivity surely is to eat and be eaten. Perhaps each molecule, each cell, is inspirited. Perhaps, to reiterate points from Chapter 2 in a new light, this is what permits intelligence to emerge through cell communication and signaling (Trewavas 2005, 414) and immune cells to remember pathogens and call on these memories in the future (Pollan 2013, 92).

THE TRANSITIVITY OF EATING AND THE VANISHING CASE FOR VEGETARIANISM

In mathematics and logic, if a bears some relation R to b, and b bears the same relation R to c, then a bears the relation R to c as well. For instance, if a equals b and b equals c, then a equals c. Likewise, if we are who we eat and our food is who our food eats, then we are who our food eats, too.[6] This is what I mean when I say that eating is transitive.

Assuming that I have provided sufficient evidence to substantiate that we are who we eat (with the acknowledgement that I have more to say on the matter when I return to the subject of ancestor veneration), this transitive relation holds. It is uncontroversial, according to Fox, that to eat animals likewise involves eating plants that the animals consumed (Fox 1999, 156). We are the People of the (Genetically Modified) Corn, after all. Maybe because the people of our culture are accustomed to thinking of ourselves as residing at the top of a food hierarchy rather than as enmeshed in an intricately laced food heterarchy, it is easier for us to think of the transitive character of eating in this way: as animals feeding on plants and as us then feeding on animals. But with the support of microbes in the soil, plants feed on the residuum of decayed bodies, including both plants and animals. They are indiscriminate eaters in this regard. This means that you and I are, too, even if we heretofore have assumed that we subsist exclusively on plant life. Let me spell this out in blunt transitive terms. Vegetarians are who plants eat, and plants eat other plants as well as animals. So vegetarians eat animals as well as plants. This is so no matter who constitutes the final strand in the food web that leads to our mouths.

Indeed, what else is rich, black soil but the composted remains of the dead (Jensen and McBay 2009, 38)? Consider, for example, the role that salmon play in the food web in the Pacific Northwest. Alex Shoumatoff notes that bears routinely carry salmon they have caught into the forest, "where the rotting carcasses, rich in nitrogen, fertilize the soil. The nitrogen permeates the trees and the flowering plants; even the snails and slugs get it. The sea feeds the forest, and the bears are the bearers of these nutritious infusions" (2015). It is not as if the soil acts as some sort of inert cleanser of animal debris. "One tablespoon of soil contains more than one million living organisms, and, yes, every one of them is eating," Lierre Keith remarks. "Soil isn't just dirt. A square meter of topsoil can contain a thousand different *species* of animals" (2009, 18; see also Mollison 1988, 205). These creatures intermingle with humus, which is a combination of humic acid and polysaccharides. "No one knows how humic acid forms, but once formed it acts like a living substance," Buhner states (2002, 165). Humic acid, which facilitates decomposition, does not discriminate between dead plants and dead animals. Neither can we. Nor does the fact that our digestive system is internal to our bodies while that of plants is external to their bodies strike me as morally relevant. What difference should it make where our food is broken down if what we are primarily concerned about is who constitutes the source of our food?

It also makes little sense to suggest that if we cannot be vegetarians, then we all must be omnivores. The distinction itself breaks down, since every member of the community of life is "one thing"—not genetically but instead in the sense that we all are equally animated by the fire of life and we all are constituted by the same nutrient cycle. Again, if we are so moved, we can distinguish between the last strand in the food web leading to our mouths and all strands that came before. But what do we gain from doing so if this leads us to deny the transitive relations that constitute the community of life? It is far better to identify what we lose: namely, a wonderful opportunity to begin dismantling our ecocidal culture's core axioms.

From an animistic perspective, the transitivity of eating is probably neither particularly troubling nor revelatory. Indeed, I suspect that it is very old news. Consider how important it is, according to animists, to maintain care-sensitive and land-sensitive relationships with who we make our food. Animists attend to the fact that there are better and worse ways to die and to kill. This gives them every reason to reject the perpetuation of both flesh factories and flesh fields. Consider also that the firm taxonomic distinction between the animal and plant kingdoms is a relatively recent formulation by the people of our culture. Contemporary transgenic research has confounded this taxonomic distinction once again. Perhaps, then, we should take heed of the fact that it is more typical for animists to speak of "deep-rooted people" or "leaf-headed standing people" rather than plant people—or "four-legged, crawling, and winged people" rather than animal people (Justice 2011, 52–53 and 63). What is most interesting about these various people from an animistic perspective is what they share, precisely that *they are all people*, not what ostensibly divides them.

More significantly, Melissa Nelson ties ancestor veneration directly to the transitive relationship that eating embodies. As she remarks, "The bones of our ancestors have become the soil, the soil grows our food, the food nourishes our bodies, and we become one, literally and metaphorically, with our homelands and territories" (2008a, 10). Stated otherwise, she adds,

> "We are what we eat," the old saying goes. But we are also *where* we eat. The plants and animals that we consume become our bodies. Our food literally becomes our flesh and our flesh gives shape to our minds and spirits. After we die, our flesh then becomes the earth, the "environment," which grows food, and the whole cycle flows all over again. This is a primal and sacred cycle of birth, growth, and death, and regeneration that most of us now take completely for granted. This "nutrient cycling," as western scientists call it, is the basis of our production, reproduction, and regeneration. (2008b, 180)

In the same vein, Jack Forbes states forthrightly that "The surface of our mother is largely composed of the transformed bodies of our relatives who have been dying for millions of years. 'Soil fertility' is, in large part, nothing but a measure of the extent to which a particular bit of ground is saturated with our dead ancestors and relatives. Death, then, is a necessary part of life" (2008, 10–11).

The health of our landbase begins with dead bodies, the detritus of live bodies, and excreted waste. It depends on death, decomposition, and decay. "That's how it has always been, since the beginning of life," state Jensen and McBay. "I eat you, the soil eats me, everyone eats the soil, you eat me, the soil eats you" (2009, 5). Whether you are a plant or an animal is immaterial. The fact that both animals and plants are able to digest whatever we eat only because we are assisted by bacteria solidifies the point. Our interdependence—our mutual insparkedness—is complete.

It is worth emphasizing, though, that being interdependent does not imply that we potentially have a problem with individuation—that we cannot possibly determine where our food begins and we end. I have already accounted for genetic diversity. And Doug Brown (2009) argues that while some animists are pantheists, many others are not. Sacredness may be immanent in the cosmos, as we have seen with postmaterialism, but this does imply that everything is subsumed in an all-encompassing source of divinity. We can count each lifeform as sacred in its own right. Harvey seems to be partial to this stance. So am I.

Moreover, we should remind ourselves that from an animistic perspective, we are all food *and* more than food. Individuals' needs and interests matter. So do the needs and interests of the landbase. Without the landbase, there would be no individuals—no us. But the primacy of the landbase does not somehow invalidate individuals' integrity. I certainly am not advocating crass utilitarianism or a reductive instrumentalism whereby we can do to and with individuals what we will for the sake of the greater good. Yes, the primacy of the landbase and respect for individuals' integrity can be in tension. But I urge you to recall Harvey's contention that this is one of the "messy realities of a shared world" (2006, 117). Agendas invariably collide. Messiness matters, which supports the proposition that there is no problem of individuation after all.

Rejecting conventional agriculture in all its forms should make the transitivity of eating easier to accept. I hope it does anyway. If you are at all like me, doing all that we can to avoid being implicated in the atrocities of flesh factories and flesh fields also should help soothe the cognitive and emotional dissonance that you may experience if my conclusions do lead you to rethink your dietary commitments. If instead you are unmoved by what I have said, so be it. For my part, I am overwhelmed by what my prior commitments led me *not* to see—what they led me to miss seeing, to neglect, and to overlook at my own folly. I can find no other way to put it.

THE FALLACY OF SEMANTIC INTRANSITIVITY

But what if omnivorism has such a radically different and much more malignant connotation than vegetarianism does? What if eating animals comes with moral baggage that eating plants might help us to unload? Were this the case, the distinction between omnivorism and vegetarianism would matter. And it is indeed the case, Fox argues. This means that eating turns out to be semantically intransitive, on his account (although this is my term, not his). Who constitutes the strand of the food web that leads to our mouths does matter because of the constellation of meanings we attach to the constituent of that final strand.

The Meaning of Omnivorism

Meat eating is most directly identifiable with the image of "man as hunter," Fox proclaims. To be a meat eater is to be a "real man," which entails being "aggressive, warlike, and predatory" (1999, 25). Ortega y Gasset (1985) provides what Fox refers to as a "classic statement of this view" (1999, 191, n. 5), a view that is starkly phallocentric. Meat eating is quintessentially masculine. It goes hand in hand with, and perhaps provides the most fitting reflection of, the idea that both women and the world are fit subjects for consumption and control. To be a real man in this sense is to display that one has the power to domesticate, to be lord and master of, one's domain (Adams 1990 and Kheel 2004).

Fox quotes Nick Fiddes, who maintains that this view of meat eating reflects "the ideology governing our species' relationship to nature" (1991, 26). Namely, consumption of dead animals exemplifies human mastery (ibid., 173). It reinforces our image of ourselves as being at the top of a food hierarchy, and it corroborates that the world and all that it contains is ours—*man's*, which really means *men's*—to conquer and subdue. (Recall here the statements by Descartes and Bacon that I cited in Chapter 1.) To be a meat eater is to confirm, states Fox, that we are "the dominant species at the same time as we satisfy our basic need for food. We also, to a degree at least, assume the right to wield power over life and death and to annihilate the alien other" (1999, 26). Animals are instruments for man's (correct that, *men's*) use. Consuming them constitutes something of a ritual sacrifice to our own quasi-divinity.

This is not merely a contemporary phenomenon, mind you. Fox cites Colin Spencer, who highlights that even in many early tribal societies, sharing meat with guests operates "as a token of the group itself, of its identity, unity, and power" (1995, 180).[7] People in early tribal societies may have been far more reliant on gathering than on hunting. Hunting yields were unreliable compared to gathering yields, and women traditionally bore responsibility for gathering. "Women were therefore more characteristic guardians of the food economy in early societies than were men," Fox remarks (see also Bell 2001, 54f.). But it was sharing meat that was valorized. And today, the central role women traditionally played in food economies is regularly downplayed, except perhaps in some anthropological circles.

Freeing ourselves from the "man the hunter" view of ourselves "permits us to look more critically at food symbolism," Fox avers. Exploding this view requires "a greater degree of objectivity. We can even construct new ways of thinking about ourselves and our relations with others as a result" (ibid.). "Equally importantly," according to Fox, "our species' self-identity need no longer rest upon the systematic infliction of suffering and death on sentient nonhumans" (1999, 26). Indeed, embracing vegetarianism reflects nothing less than an "evolving collective conscience" (ibid., 29).

The Meaning of Vegetarianism

Mary Midgley states forthrightly that "the symbolism of meat-eating is never neutral. To himself, the meat eater seems to be eating life. To the vegetarian, he seems to be eating death. There is a kind of gestalt shift between the two positions which makes it hard to change, and hard to raise questions on the matter at all without becoming embattled" (1983, 27). For vegetarians, meat eating does not connote power and manliness. It instead represents the lamentable and unnecessary loss of sentient life. Fox acknowledges that "we cannot escape participating in the process that produces plants' deaths if we are vegetarians. But the latter [plant life] lacks the literal as well as the symbolic significance of the former [animal life], given that we are not implicated in acts of terminating conscious, sentient life—life that we endow with value of a different and higher order than we do vegetative life" (1999, 33). Because vegetarians do no harm to sentient beings, according to Fox, their dietary choices act as a counterweight to the oppression of women and of animals that omnivorism embodies. Altering what we eat—and it is *what* we eat for Fox—in this way surely is not the only solution to the abiding problems of phallocentrism and anthropocentrism. But it provides at least one salient "opportunity to help weaken other links in the chain of oppressions" (ibid., 37). If nothing else, it is a very good place to start.

Weakening these links is made possible in part by the fact that vegetarianism connotes that we are in a good position and are interested in developing a stronger epistemological and emotional connection with our food. It heightens our sense of the ecological consequences of food production, gives us a sense of responsibility to end the abuse of our world, and helps us develop "a deeper understanding of our dependence on nature" (ibid., 36–37). Consider, for example, the embrace of organic gardening among vegetarians, Fox notes. Consider, moreover, how this can give us a better appreciation of the food we produce. To overlook these phenomena with claims of transitivity masks the violence of omnivorism and the healing power of vegetarianism. They symbolize radically different stances toward what we make our food, Fox concludes, which we overlook at our own peril.

Who Are "We"?

To be clear, Fox's goal of eradicating phallocentrism and anthropocentrism from our culture is a worthy one. I suggest, though, that the way he frames the distinction between vegetarianism and omnivorism gives us a skewed view of what they symbolize, which ends up doing a disservice to his cause. I see two main problems with Fox's defense of semantic intransitivity. First, he identifies the "man as hunter" view as a human problem, while I suggest that it is instead best regarded as a problem endemic to the people of our culture. Second, his understanding of what vegetarianism means reflects the same sorts of assumptions maintained by sentientists. So his understanding of vegetarianism is burdened by the very misconceptions and prejudices that we dealt with in Chapter 2.

Who are we? Who are the "we" who are predisposed to accept the "man as hunter" view, according to Fox? Who are the "we" who discount the role of women in the food economies of our ancestors? Indeed, who are the "we" who profess that plants are not sentient beings and are therefore deserving of being endowed with lower moral standing than animals? Fox endorses Fiddes's claim, cited previously, that these stances govern "our species' relationship with nature." Fox himself is forthright that "our species' self-identity" must no longer be dependent on the notion that we have the right to inflict systematic harm on "sentient nonhumans." And Fox seems to assume that it is common knowledge—that it should be taken as a matter of course—that plants are not sentient and are not deserving of equal moral standing.

But Quinn provides some welcome news that Fox would do well to heed. "*We are not humanity*," Quinn declares. "Can you feel the liberation in those words? Try them out. Go ahead. Just whisper them to yourselves: *We . . . are not . . . humanity*" (1996, 285). The meaning of omnivorism with which Fox is concerned is not universally endorsed. It reflects not a *human* problem with how we view our food but a *cultural* problem instead. In acknowledging this, I do not wish to whitewash the ways in which indigenous peoples, too, can be complicit in ecological destruction and harm to fellow members of the community of life. Instead, I highlight that Fox's defense of semantic intransitivity is supported in part by a common error made by the people of our culture: forgetting that we are representatives of just one culture rather than humanity as a whole. The way we think and act is not necessarily universal.

What exactly makes this news welcome? Were we challenging a human problem in combating the "man as hunter" meaning of omnivorism, I dare say it is highly unlikely that our efforts would be successful. The problem would be too big, too widespread, perhaps even too deeply ingrained in our sense of self.

"But it isn't humanity that needs changing, it's just . . . us,"
Quinn contends (ibid., 286).

Again, I do not say this to belittle Fox's concerns. On the
one hand, better framing of the problem can help us to combat phallocentrism and anthropocentrism more effectively. On
the other hand, it is fitting and proper to acknowledge that
omnivorism does not have a single, universally applicable meaning. What meat eating in general symbolizes is not necessarily
phallocentric and anthropocentric. This is what it represents in
our culture, not in *every* culture. Some ways of understanding
what meat eating signifies work much better than others—for
people and for the planet. Look back at the Ihalmiut example
and judge for yourself.

Let us now proceed to the second problem, that Fox's understanding of vegetarianism bears the hallmarks—and attending
burdens—of sentientism. His sense of what vegetarianism symbolizes is misconceived. As a result, his argument for semantic
intransitivity fails. We should not rush to conclude that what
meat eating means is incommensurable with what plant eating
means so long as our understanding of one or both is errant.

Fox's explanation of the meaning of vegetarianism hinges
on two crucial claims. First, by looking with "a greater degree
of objectivity" at what our food symbolizes, we put ourselves
in a much better position to develop a self-identity that is no
longer supported by harming or killing "sentient nonhumans."
Second, vegetarianism heightens our sense of the ecological
consequences of food production, so being a vegetarian makes
us more ecologically conscious.

With respect to the first claim, Fox's understanding of what
vegetarianism symbolizes is based on a flawed understanding
not only of plants but also of the food web and our place in
it. So he ends up seeing differences between eating meat and
eating plants that are not really there. This alone is fatal for his
defense of semantic intransitivity. Plants *are* sentient. More than
that, they arguably are *people*. Based on either criterion, surely
they are deserving of equal moral standing. I trust that I do not
need to rehash my argument for these propositions. Suffice it

to say that vegetarians countenance the killing of sentient other-than human beings, too. It is not only omnivores who do so. The problem we face with respect to making sentient beings our food hereby is not *that* we kill but *how* we kill and how we treat the beings who we make our food before they become our food.

For one thing, violence does not equal aggression, Jensen proclaims: "There is the necessary violence of survival, the killing of one's food, whether the food is lettuce, onion, duck, or deer. Then there's the senseless forms of violence so often perpetrated by our culture: child abuse, rape, military or economic genocide, factory farms, industrial forestry, commercial fishing" (2000, 35). Jensen's claim raises a number of vexing questions. Is eating really a form of violence? Do we commit violence when we eat if we are not the people who kill our food? Jensen and McBay suggest that it is not possible for individuals to live sustainably in a structurally unsustainable culture (2009, 60). In the same vein, is it possible to avoid being complicit in aggression in a culture in which it is endemic? Each of these questions requires attention that I am not equipped to offer in this forum.

I agree with Jensen, though. I cannot help but agree with Quinn, too, when he insists that part of what vegetarianism has come to symbolize in our culture is that one has the privilege of feeling morally superior if one refuses to eat animals. This is kingdomist through and through. In effect, Quinn concludes, vegetarians assign "a greater sacredness to members of their own kingdom (the animal kingdom) than to the plant kingdom" (1994, 165). Forgive the polemics, but I find it difficult to avoid seeing this same dynamic at work in Fox's defense of semantic intransitivity. Surely we can challenge phallocentrism and anthropocentrism without recourse to professions of moral superiority, particularly professions that are so poorly supported.

What of Fox's understanding of the food web and our place in it? "[B]y eating at the bottom of the food pyramid," he states, "vegetarians can feel that they are doing something good for the world around them. [. . .] This can enable them to

experience a sense of 'rootedness' or 'at-homeness' in nature"
(1999, 37). But higher or lower with respect to what—or who,
really? Recall that animals are valued over plants, according to
Fox, because the former are of a "higher order" than the latter. I
get the distinct impression from him that it is we who are at this
pyramid's apex, which defies his rejection of anthropocentrism. I
do not deny that humans are special. But so is every living being,
without exception. Indeed, we all are both exceptional and unex-
ceptional, and we should not want it any other way.

My bigger concern is with Fox's assumption that there are
higher and lower stations among whom we make our food in
the first place. Fox here appropriates Aristotle's *scala naturae*,
which we already know belies the breathtaking complexity
and continuity of life. When we substitute Fox's food pyramid
with a food web, this is easier to see. Those who are deemed
"lower" by Fox must eat, too, of course. What do they eat—
those who are lower than them? Who is that? And do those
who are lower than them eat what is lower than them in turn?
No, what we have is a food heterarchy, not a food hierarchy.
I like the way Keith puts this: "everything is eating and then
being eaten, and through it all life endures. There is no hierarchy, only hunger. And it's through hunger that we participate
in the cosmos, in an endless cycle of life, death, and regeneration" (2009, 72).

On its face, Keith's remarks bear some similarity to a passage that Fox cites from Mark Mathew Braunstein: "If we are
what we eat then we must ask: What do we eat? And from
where does what we eat come? Our food either comes directly
from plants, or from animals who come from plants. In either
case we indirectly come from what plants directly come from,
so we are really eating the sun, the earth, the stars, and the
moon" (1993, 48). But there are critical differences between
the points that Keith and Braunstein, respectively, make.
Fox emphasizes that Braunstein's considerations can help us
develop "a much more profound insight into where we stand
in the universe, an insight that may appropriately evoke in us
a sense of gratitude and humility" (1999, 38). I fully support

expressing gratitude and humility for whom we make our food. Surely Keith does as well. But I think she and I take issue with the idea that what Braunstein offers us is profound insight. Yes, plants feed animals. But animals feed plants, too. "Grass and grazers need each other as much as predators and prey," Keith concludes. "We aren't exploiting each other by eating. We are only taking turns" (2009, 8).[8] If she is willing to add that eating does not *necessarily* involve exploitation, then she and I are on the same page.

Let us move on to Fox's second claim: that vegetarianism enhances our sense of the ecological consequences of food production, thereby facilitating our ecological consciousness. My reply is brief. Fox's considerations may support our understanding of the stark ecological harm done by the animal-industrial complex, by flesh factories. But I dare say that it does little to help us see how ecologically harmful flesh fields are—except to the extent that they support the perpetuation of flesh factories. Indeed, as I mentioned in the previous chapter, Plumwood contends that to be a vegan, one must depend on an ecologically catastrophic global food economy. She may be wrong, as I suggested. But my point is that vegetarians and vegans do not often countenance either the ecological ills of a global food economy dominated by flesh fields or their own dependence on the perpetuation of these ills to maintain their dietary habits. On this front, too, they see intransitivity between eating meat and eating plants where there is none.

So the community of life, of which we inextricably are part, is a closed-loop system. We are who we eat. We also are who we eat eats via the transitivity of eating. Because who we eat eats both animal life and plant life, we cannot be vegetarians. But neither are we omnivores, since we—us, our food, and our ancestors—are all or have all at some point been animated by the same fire of life. We are all "one thing," Quinn emphasizes, so there is no "omni" to "vore." The very distinction between vegetarianism and omnivorism derives from a view of ourselves and our place in the world that is rooted in the axioms of our ecocidal culture.

Look at the assumptions underlying Fox's defense of semantic intransitivity. Consider how they reflect the axioms of our
culture. The meaning of meat eating poses a human problem,
Fox insists, not a cultural problem. Perhaps this is because we,
the people of our culture, are the quintessence of humanity. We
reside at the apex of the food pyramid, he asserts, which indicates that the world was in some sense made for us. We sentient
beings belong to a separate, higher order, than all other beings
on his account. And vegetarians' sense of the moral superiority
of their eating habits, which is embodied in Fox's defense of
semantic intransitivity, may well presuppose that these habits
must be ecologically friendly. From this perspective, the earth
was built to keep on giving indefinitely, at least so long as one
is a vegetarian. No, no, no, and no. Let us bid farewell to the
case for vegetarianism. Let us permit it to vanish. This case has
outlived whatever usefulness it once may have had. It no longer
works, if it ever did.

CHAPTER 5

Two Objections, One Accommodation

In this penultimate chapter, I address two anticipated objections to my argument. According to the first, I have overlooked circumstances that still make an ecological defense of vegetarianism salient, whether or not we specify it in moral terms. We can accept the basic merits of the contextualist defense of vegetarianism and still argue that, ceteris paribus, vegetarianism is ecologically preferable to omnivorism. Relying primarily on the work of Simon Fairlie, I examine the evidence offered on behalf of this objection and suggest that matters are considerably more complicated than those who offer it let on. My reply does not nullify its warrant outright. I suggest, however, that objectors have considerably more work to do to make their case.

The second objection alludes to concerns voiced by Mark Boyle, the "Moneyless Man" (2010), who asserts that questions of the ecological viability of dietary choices must be circumscribed by a wider discussion of human overpopulation and overconsumption. It makes little sense to offer grand claims about who and how we eat without accounting for the fact that our landbase almost certainly needs fewer of us consuming far less of it and the myriad kinds of people who inhabit it. Boyle is right. Human overpopulation and overconsumption must be regarded primarily as ecological problems that have ecological solutions rather than as they are generally categorized: as social and technological problems that have little or nothing to do with the fact that we are eating our way through the rest of the community of life. Any discussion of the merits of our respective dietary proclivities must be framed within these wider ecological terms. The number of human eaters and the amount we consume matters at least in equal measure to who we eat and how we eat who we eat.

After responding to these anticipated objections, I close the chapter by suggesting that there is still a way—from an animistic perspective—to honor certain aspects of what motivates many people to be ethical vegetarians. Despite the shortcomings of both expansionary sentientism and a care-sensitive ecological contextualism as defenses of vegetarianism per se, the need to account for wider concerns associated with human overpopulation and overconsumption, and even the vanishing case for vegetarianism, it is possible to specify conditions in which we can accommodate the vegetarian impulse. Simply put, there is an animistically inspired way to think about the appeal of maintaining special obligations to animals—as vegetarians are wont to do—without denying plants their due. Vegetarianism may be morally indefensible and ontologically illusory, but perhaps there are still aspects of what motivates people to embrace ethical vegetarianism that are important to acknowledge.

ALL THINGS BEING EQUAL . . .

We saw in Chapter 2 that the extremity of the ecological harm done by flesh factories seems to weigh heavily in favor of embracing vegetarianism. The animal-industrial complex as a whole is accountable for widespread deforestation, soil erosion, loss of soil fertility, desertification, and exorbitant water usage and water pollution (Taylor 1986, Durning and Brough 1991, Rifkin 1992, 201, and Pluhar 2004). It also contributes to biodiversity loss and global climate change (Regenstein 1985, Schleifer 1985, 68, Amato and Partridge 1989, 19, Hill 1996, 111, and Fox 2000). Even if the maintenance of a vegetarian diet relies on flesh fields and a global food economy, it surely is not as ecologically harmful as an omnivore diet that depends on flesh factories. This is the conclusion that Singer draws in his attempt to deflect concerns about plant sentience.

But if we accept the landbase as primary and hereby reject conventional farming in all its forms, what we need to know is how the local and organic production of plants for food is likely to measure up *in our specific context* against the local and organic production of animals for food. It is obviously impossible to provide a comparative analysis for every feasible context. Instead, I draw on Fairlie's meticulous analysis of the United Kingdom, from which it is possible to draw some general conclusions.

The primary environmental objection raised by vegetarians to eating animals is that this practice constitutes an inefficient means of obtaining food. Far more lives are destroyed and far more water and soil nutrients are used to subsist on animals than to subsist on plants. Objectors also raise concerns about carbon and methane emissions that are produced by grazing livestock. Yet Fairlie challenges some oft-cited statistics regarding the degree of inefficiency of meat production. The claims that the feed conversion ratio is something like ten units of inputs for every one unit of outputs (Rifkin 1992, 160, Smil 2001, 165, Singer 2002, 165, and Whitefield 2004, 258), that 20,000 liters of water or more are required to produce one kilogram of beef (Amato and Partridge 1989, 19, Hill 1996, 111, Nelson 2004, and Robbins n.d.), and that livestock are responsible for 18 percent of global greenhouse gas emissions (Food and Agriculture Organization of the United Nations 2006) do not add up. Indeed, they do not even come close to adding up, Fairlie argues. Nor is raising livestock in and of itself damaging to water and land (Fairlie 2010, 78, and Savory 2013).

These statistics apply in particular to cattle feedlots, which Fairlie calls "one of the biggest ecological cock-ups in modern history" (2010, 9). The cattle industry in the United States is the main culprit. It also is an aberration in terms of international norms and practices. Throughout much of the rest of the world, cattle are grass fed. This is as it should be, since their multiple stomachs are designed to extract carbohydrates and protein from fibrous grasses. By contrast, Monbiot notes, the American cattle industry "pumps grain and forage from irrigated pasture lands into the farm animal species least able to process them efficiently, to produce beef fatty enough for hamburger production. Cattle are excellent converters of grass but terrible converters of concentrated feed," which is their main source of sustenance in factory farms (2010). If we account for the fact that cattle throughout much of the world subsist on grasses instead of grains, the feed-to-food ratio is perhaps closer to three to one, Fairlie contends. If we also factor in byproducts of meat production—including leather, manure, soap, insulin, and blood plasma—and the fact that the production of pigs, poultry, and fish is more efficient than the production of cattle, the ratio is reduced even further (2010, 19ff.).[1]

Industrial cattle production in particular is extremely wasteful of water. But there is only one possible way to conclude that a single kilogram of beef requires the input of 20,000 liters or more of water, Fairlie declares (2010, 64). This is to assume that every drop of precipitation that falls over the course of a year on land either that cattle occupy or that produce feed for them is lost in their production. None of this water is expelled back into the soil. But, as Keith remarks, "animals aren't ever-expanding water balloons. For a steer, almost all of that water will be returned in the form of urine and feces laden with nutrients and bacteria, value-laden as it were, to the land that needs it" (2009, 103). To be sure, Monbiot states, "A ridiculous amount of fossil water is used to feed cattle on irrigated crops in California, but this is a stark exception" (2010). By contrast, the average water usage for grass-fed free-range cattle who do not graze on arid land and who are slaughtered in small-scale slaughterhouses is perhaps closer to the 200 kilograms contained in their bodies upon death (Fairlie 2010, 67). Let me be clear. That is 200 kilograms for the 220 kilograms of beef their bodies yield on average, or 0.9 liters of water per kilogram of beef.

Lastly, the Food and Agriculture Organization's (FAO's) conclusion that meat production is responsible for 18 percent of global greenhouse gas emissions relies on several dubious claims. The main reason the FAO's figure is so high is that it includes emissions associated with Amazon deforestation. Even if the cattle industry alone were responsible for these emissions, Fairlie notes that they reflect an expansion of this industry rather than cattle production itself. But these emissions should not be attributed to cattle production in the first place, since Amazon deforestation has been driven mostly by logging and land speculation, not by ranching (Fairlie, Young, and Thomas 2008).[2]

Fairlie's statistics come thick and fast throughout his presentation. I advocate having a look at them for yourself, since, as he himself states, "different writers come up with different figures, which are often a reflection of their ideological position" (2010, 15). I set these matters aside, though, so that I can proceed to my main point.

We must assume, notes Fairlie, that ours is a society that will soon be "in a state of energy descent, with increasing dependence upon renewable resources, more waste cycling, and (consequently) a localized economy which is more integrated with natural processes" (2010, 100). We must also keep in mind that everything we eat has an ecological footprint (Fairlie, Young, and Thomas 2008, 19). Yes, ruminants have a large footprint. But plant foods transported long distances, most notably when frozen or grown in heated greenhouses, tend to be less ecologically sound than locally and organically raising animals for food. Some vegetarian mainstays—tropical fruits, almonds, hothouse tomatoes, cheese, chocolate, coffee and tea, cereals, herbs and spices—have especially large footprints.[3]

So let us begin by stipulating that a comparison of vegetarianism and omnivorism that fits with the wider concerns I have outlined in this work must weigh the most ecologically sound approach to the former against the most ecologically sound approach to the latter. This is the most fitting way to determine whether vegetarianism is, ceteris paribus, preferable to omnivorism, hence whether an ecological defense of vegetarianism is indeed viable. To make such a determination, Fairlie stipulates that we must compare the *amount of land* required sustainably to raise animals for food to the amount of land required sustainably to raise plants for food of the same nutritional value.[4] So let us proceed by assessing how what Fairlie calls "livestock permaculture" measures up against what he calls "vegan permaculture."

At root, remarks Patrick Whitefield, permaculture "means taking natural ecosystems as a model for our own human habitats. Natural ecosystems are, almost by definition, sustainable, and if we can understand how they work we can use that understanding to make our own lives more sustainable" (2004, 3). Permacultures are designed to be ecologically diverse, stable, and resilient. Diversity, stability, and resilience work with rather than against the landbase. As Buhner states, "The larger the number of plants with diverse chemistries that occupy the largest number of ecosystem functional categories, the more vital and healthier the ecosystem. This is because no year is ever the same as the previous year; environmental conditions are always different. Local habitats are always shifting in response to changing conditions. A plant producing major contributions to habitat need one year may not play the same role when environmental factors change. This is why a large range of plants in a local community is necessary: to give the system maximal response ability" (2002, 185). While they can be high yielding, permacultures run on contemporary sunlight rather than on the "ancient sunlight" embodied by fossil fuels (Hartmann 2013, 8). So they are low-energy, self-perpetuating systems that serve as ecologically sustainable homes for humans, other animals, and plants alike (Mollison 1988).

Permaculture is best facilitated by perennial polyculture, or what Wes Jackson calls "ecosystem agriculture" (2010, 479). Self-seeding annuals prove especially beneficial for rebuilding disrupted ecosystems. But stable ecosystems are largely populated by perennials who can live for many years, and they have roots that extend deep enough into the soil that they draw nutrients from rock material well below the surface. This builds soil. Herbaceous understories, vining plants, trees, and fungi are grown together to better anchor existing soil, which virtually

eliminates erosion and runoff (Keith 2009, 35). Perennial polycultures likewise require no tillage, provide most or all of their
own fertility, handle pathogens and pests without petrochemicals, and sequester carbon. Unlike annuals, perennials must
devote a portion of their energy to maintaining and expanding
their inedible infrastructure (Fairlie 2010, 245). But perennials
can yield food over an entire growing season, since they start
out at the beginning of the season as mature plants (Picasso et
al. 2008 and Yanming et al. 2011). And species complementarity increases food production over time (Cardinale et al. 2007).
Organizations like the Land Institute, based in Kansas, are
even experimenting with growing high-yield perennial grains
in order to redevelop prairie land (Jackson 2010 and Pimentel
et al. 2012).

The question that we must answer is this: all things being
equal, how do perennial polycultures in which livestock is raised
measure up against vegan perennial polycultures in terms of
enhancing "the richness, the diversity, and the interconnected-
ness of the entire system" of which they are part (Fairlie 2010, 43)?
Under what Fairlie calls "default" circumstances, livestock
activity, nourishment, and waste production can play a complementary role with respect to cultivating plants. Livestock
permaculture involves a great reduction in the use of animals
for food compared to current levels of use because no landbase
can sustainably handle habitation by livestock at current levels.
Moreover, rather than being grain fed, "the main role of animals is to turn vegetation that cannot be eaten or economically
harvested by humans into useful goods and services," states
Fairlie (2010, 36). Incorporating pigs into a perennial polyculture also provides repositories for food waste as well as for
grain surpluses that result from bumper crops. Livestock permaculture thus operates as an integrated system for cultivating
plants in which nonarable land and plant materials unsuitable
for human consumption are used to raise animals for food and
other products. Keeping livestock also can prove invaluable for
transporting nutrients from pastures to cultivated land via the

practice of folding. Livestock manure can be used as well for fuel and as a building material (ibid., 29).

By comparison, vegan permaculture is stockless. It does not involve raising animals for food. Nor do vegan permaculturalists make use of animal manure or keep animals to process food waste and grain surpluses. It should go without saying that it is impossible to remove all animals from the land without making it utterly uninhabitable. It is also impossible to prevent them from dying, decaying, and decomposing to return to the soil. But I leave this matter aside for present purposes. I also set aside otherwise worthy concerns about the unintentional killing of birds, vermin, worms, and other animals in the process of harvesting plants, particularly since this is much more likely to occur via conventional planting and harvesting. The main point that I want to highlight is that vegan permaculture may be viable in the United Kingdom, according to Fairlie, but success with it faces obstacles. Notably, it incorporates "green manure," or legume crops that act as plant-based sources of nitrogen, to enhance soil fertility. And whereas Fairlie calculates that livestock permaculture in the United Kingdom requires 1 hectare of arable land plus 0.8 hectares of pasture to supply 7.5 human people with food, fiber for construction, and fuel, he calculates that vegan permaculture requires 1 hectare of arable land to supply 8 human people (ibid., 98f.).

Based on the amount of land used, vegan permaculture clearly comes out ahead. It mainly economizes on pastureland (ibid., 212), which can be rewilded and perhaps even used for wildcrafting. But there are other matters to consider. Fairlie contends that 33 percent of arable land must be set aside by vegan growers to cultivate green manure, although this can be reduced via the use of cover crops, oil residue, or the importation of compost (ibid., 103). This means that vegan growers may have at their disposal less arable land on which to grow food than livestock growers do. Moreover, vegan growers are less able to deal with harvest fluctuations, since they cannot rely on animals to provide a food buffer when grain reserves run short or a repository to prevent bumper crops from going to waste (ibid., 114f.). It is worth noting, though, that numerous other crops—including pulses and nuts—easily can be stored for future use.

Fairlie worries that vegan growers also may run into problems when it comes to maintaining the soil's phosphorous levels. Phosphates stay constant when everyone who lives on a landbase on which vegan permaculture is practiced also dies there without ever traveling long distances. Problems arise in particular when crops are exported or are eaten without the phosphates they contain being returned to the land on which they are grown. By comparison, in livestock permaculture, animals can extract phosphorous by "grazing the hinterland" and then being folded onto arable land to deposit their waste (ibid., 102).

Finally, at least within the United Kingdom, the production of fats presents something of a problem. Until the twentieth century, most fat was of animal origin in places like the United Kingdom that are largely unsuitable for producing pressed oils. The further away from the equator one goes, the more difficult it is to produce pressed oil locally. As Fairlie remarks, "Animals grow a layer of fat to keep warm, whereas plants contain oil to stop their seeds from drying out; therefore animals fats are found mostly in cold climates, while vegetable oils are abundant in warm climates. There are exceptions such as linseed and rapeseed, but until recently there were regarded as inedible and used only for industrial purposes. In the north, animal fats are local foods—the rise of vegetable oil is a symptom of globalization" (ibid., 56; see also Nonhebel 2004).

All told, then, the ecological comparison of livestock permaculture and vegan permaculture is not easily adjudicated. In some warmer climes, the ecological case for veganism may well be stronger. In some northern climes, it is considerably weaker. Indeed, although Fairlie does not account for them, some locales also may be suitable for vegetarian and semivegetarian permacultures, both of which offer intermediary approaches to food production. Other locales may be flexible enough to allow for livestock permaculture, vegan permaculture, and intermediary variations to exist side by side "as long as boundaries are drawn and fences maintained," Fairlie states (2010, 103).

Vegan and vegetarian growers may even find livestock growers to be good neighbors, since livestock permaculture necessitates a great reduction in the eating of animals. Moreover, insofar as livestock growers keep fewer animals, they are likely to lavish considerably greater attention—and perhaps affection—on those animals (ibid., 277). So the ecological distinguishability between these various approaches to permaculture ultimately may not be as pronounced as affected parties generally suggest. And they may well share the desire to maintain a care-sensitive relationship with their food, too.

HUMAN OVERPOPULATION AND OVERCONSUMPTION

Commenting on Fairlie's considerations, Boyle states forthrightly that "I find this entire approach [to be] in danger of reducing all life to a carbon footprint equation [. . .], and it doesn't get to the real heart of the matter" (2011). More relevant, argues Boyle, are problems associated with human overpopulation and overconsumption. The former is most acute in areas of the world heavily populated by the poorest of the global poor. The latter phenomenon is especially problematic in industrialized nations, particularly in the United States, which bear almost total responsibility for global climate change, an emerging mass extinction event, and, of course, the transfer of wealth from shadow places to beautiful places. The former phenomenon also may be attributable to the actions of industrialized nations even if it is most rampant in global shadow places.

Boyle is right. These phenomena do not invalidate the debate over our eating practices. But any discussion of who we eat and how we eat who we eat must be considered within a wider context of how many of us eaters there are and how much we eat.[5]

The most common approach to addressing overpopulation and overconsumption is to call for simultaneous social and technological revolutions. With respect to the former, John Firor and Judith Jacobsen remark that "poverty, especially in rural cultures where a child's labor is a meaningful contribution to the household economy and where children provide a couple's only financial security in old age, does indeed cry out for many children" (2002, 36; see also Commoner 1975 and Brown 2001, 223). Whereas the expense of having children outweighs its financial benefit among those who are better off, children may provide the only viable "asset" for the poorest of the global poor. For the rural global poor, the more children that parents have on hand, the more free labor they have at their disposal to work the fields and tend the animals. Children of poor urban families, particularly boys, can work instead of going to school in order to increase the family's aggregate income.[6]

In order to reduce the financial incentives among the global poor associated with having children, Firor and Jacobsen call for "a social revolution characterized by greater equity and opportunity, including better health, for the poor of all countries, but with a special emphasis on women and girls" (2002, 189; see also Brown 2001, 211, and Guillebaud and Hayes 2009). Abject poverty must be alleviated through the direct infusion of resources as well as through the facilitation of capabilities that tend to lead women to delay childbearing. Poor women and girls in particular need educational and economic outlets that provide them with the opportunity to be breadwinners, which can improve their social status—hence their control over their bodies—while also reducing the familial demand for and need to have more children. Women and girls also need effective family-planning services, including ready access to contraception (Willott 2012, 526).[7]

Taking these steps is easier said than done, of course. As Firor and Jacobsen note, "Something else drives high fertility, too. If you were to ask a woman in a high-fertility country why she has many children, she might answer as follows: 'I am nothing in my culture unless I bear children—in fact, I am nothing in my culture unless I bear sons. Sometimes the daughters keep coming, and I have to keep having babies until I've produced a son. Or two. And I was really nobody until I married, so I married young. As soon as I married, the babies started to come'" (2002, 36–37). Giving women and girls educational and employment opportunities surely can help. But women and girls are likely to take advantage of these opportunities only if they provide prestige—not just for women and girls but for the men and boys with whom they are associated—that is comparable to that which currently is gained through marriage and childbearing. This sort of cultural transformation likely cannot be, and probably should not be, imposed from without.

But short of a rapid reduction in global fertility rates or a rapid increase in human mortality, demographic projections suggest that policy changes of this sort are insufficient on their own to curb human population growth (Bradshaw and Brook 2014). They also do nothing to address overconsumption. So Firor and Jacobsen insist that a technological revolution is required alongside the sort of social revolution they describe to facilitate a dramatic drop in fossil-fuel use and per capita carbon footprints. The use of all energy and materials must become considerably more efficient, and waste basically must be eliminated (McDonough and Braungart 2002). To be sure, global climate change can be expected to make it increasingly difficult to implement technological shifts like these. But if some such shifts can take place, they may be able to mitigate the worst effects of global climate change. So even incremental success with a technological revolution can create conditions that are more conducive to further technological advancement.

I do not doubt that seeking to facilitate these kinds of social and technological transformations has considerable merit. My concern is that focusing exclusively on them obscures a more comprehensive ecological revolution that is required to reverse human overpopulation and overconsumption. Recall that the community of life is a closed loop. Its members, including us, are part of a dynamic balance that is sustained by feeding and being fed upon. If there are too may deer on a specific land-base, they eat through available food. Without available food, their population declines. This in turn leads their food to be replenished. So feeders and those who are fed upon regulate one another.[8] The same goes for us and for our sources of food,

Quinn notes. "An increase in food availability for a species means growth. A reduction means decline. Every time, ever and always. *Semper et ubique*. Without exception. Never otherwise. [. . .] More food, Growth. Less food, Decline. Count on it" (1996, 294; see also Hopfenberg and Pimentel 2001).

The community of life is currently in a period of mass extinction at the same time that the number of human beings on Earth is rapidly increasing. This can mean only one thing. We are in the process of converting biomass exclusively into human mass. As Quinn puts it, "This is what happens when we clear a piece of land of wildlife and replant it with human crops. This land was supporting a biomass comprising hundreds of thousands of species and tens of millions of individuals. Now all the productivity of that land is being turned into human mass, literally into human flesh. Every day all over the world diversity is disappearing as more and more of our planet's biomass is being turned into human mass" (1999, 113). We are preventing other species from accessing food by taking their habitats and converting those habitats into land on which we produce food exclusively for ourselves. This is what conventional agriculture is all about. In order to maintain human mass—in order for there to be seven billion or eight billion or nine billion humans on this planet—we need to go right on doing the same (Quinn 2007, 17).[9] For there to be more humans on Earth, the rest of the community of life must diminish in size. *Semper et ubique*. Always and everywhere. "It's not an accident. It's not an oversight. It's not a bit of carelessness on our part," Quinn declares (ibid., 175). What else are pests or weeds than creatures who must be eliminated or displaced because they compete for "our" food?

It should go without saying that this cannot be sustained. Quinn notes that it is as if we live on the upper levels of a skyscraper and every day descend to lower levels and knock out a steel girder or two to spruce up our accommodations above. Eventually the edifice will collapse. And so will our population if we continue on as we have. We are not somehow exempt from the fate that befalls every other species in our predicament.

It makes no difference that what our species faces is unprecedented in that the carrying capacity for the human population has exceeded what Graham Zabel (2009) calls "biomass population," or the population that is sustainable on the basis of biomass alone (see also Tverberg 2012). Carrying capacity is a measure of the energy available to support a given population.[10] A spike in human population occurred with the widespread use of coal in the 1850s, and an even bigger spike occurred with a massive increase in oil use a century later. The former improved transportation and permitted the development of more powerful farm equipment. The latter has literally been fed to the plants we grow for food, which has greatly increased agricultural output.

But this cannot last. Remaining oil either will run out or, preferably, will be left in the ground. Unless a comparable source of energy becomes available, which appears to be unlikely, our numbers will decrease. The net energy from solar and wind power is comparably low, Zabel argues. The supply of natural gas is not sufficient to offset oil. And scaling up nuclear power would require far greater energy inputs than is feasible (Friedman 2009, 260). As he states, "If we hypothesize that the Earth's population is ultimately determined by availability of energy resources, and if some of those energy resources are at or near their peak rates of production, then that may affect rates of population growth. If the correlation is strong enough, the number of people the Earth can support may also be at or near its peak" (2009). Zabel estimates that human biomass population is around one billion.[11] For our numbers to return to this once again would mark a decline of more than 85 percent with respect to our current population size. And even this may be an overestimate of the number of us that our warming world can sustain in coming years as it becomes less hospitable for us and for our food (Jamail 2013). Add to this the disastrous effects of conventional farming on soil fertility and impending water shortages (Postel 1999), and the situation looks bleak. So whether the commonly cited UN projection that there will be more than nine billion humans on

Earth by 2050 actually materializes may end up being the least of our concerns (Brown 2001, 202).[12]

I am not suggesting that we resign ourselves to the proposition that nothing can be done to make the inevitable decline in human population less precipitous. Nor must we side with neo-Malthusians like Paul Ehrlich and Garrett Hardin, who defend coercive population reduction policies and the refusal of support to the global poor (Ehrlich 1968 and Hardin 1974). We can explore ways to stabilize food availability and still distribute available food far more equitably (Sen 1994). We can explore the prospect, that is, that an ecological revolution can be driven at least in part by egalitarian political policy (Winter and Titelbaum 2013 and Weisman 2013).[13] But we must be careful with how we proceed. Humanitarian interventions are bound to fail, states Teju Cole (2012), if those who undertake them do not account for the "larger disasters"—militarization, short-sighted agricultural policies, resource extraction, government corruption, and so on—that led to the inequitable distribution of food in the first place.[14]

I suggest that the considerations offered by Plumwood, Monbiot, Fairlie, and others cited here provide helpful points of departure, since the positions they espouse are consonant with the establishment of just bases for the conversion of human mass back into biomass. Are there easy ways to carry out such an undertaking? Of course there are not. But the more pertinent issue is whether we are prepared to take steps to facilitate a relatively smooth decline in human population and consumption or whether a far more precipitous decline in both is thrust upon us.

ECOLOGICAL FLEXIBILITY AND TOTEMIC ACCOMMODATION

So let us once again acknowledge the dire need for ecological restoration. Let us recall Jensen and McBay's point, cited in the Chapter 3, that healthy land is giving land. The less healthy the land, the less it can give. Grievously wounded land is likely to die if it endures additional injury (2009, 59). Ecological restoration thus entails the reestablishment of ecological flexibility, of the capacity for the land to readily renew itself. This is the clearest sign that the land is healthy (Leopold 1949, 258, and Hoagland and Dodson 1995, 144). And self-renewal is most directly facilitated by optimizing biodiversity. I repeat Buhner's perceptive claim: "The larger the number of plants with diverse chemistries that occupy the largest number of ecosystem functional categories, the more vital and healthier the ecosystem. This is because no year is ever the same as the previous year; environmental conditions are always different. Local habitats are always shifting in response to changing conditions. A plant producing major contributions to habitat need one year may not play the same role when environmental factors change. This is why a large range of plants in a local community is necessary: to give the system maximal response ability" (2002, 185).[15] Facilitating ecological restoration becomes all the more pressing in the face of global climate change (Light 2012). Especially given the uncertainty this involves, we need landbases that tolerate not just acute climactic disruptions but also our ignorance about how to cope with changes in their needs and interests that these disruptions may engender. We need them to be able to forgive us for our errors (Jackson 2010, 480).

If ecological restoration succeeds, perhaps we also can more easily honor individuals' vegetarian impulses even on landbases not particularly well suited to vegetarianism. I make this point not to justify inaction. Nor am I trying to compensate for my brief journey into polemics in Chapter 4. Rather, I want to be respectful of certain of the intentions and emotions that motivate vegetarianism. I also want to envision how to bridge the chasm between where we are conceptually in terms of the way we relate to our food and where, with no shortage of luck and diligence, we can end up.

I take my inspiration in this regard from Barbara Kingsolver, who states the following about those of us who have come of age over the last few decades—particularly in the United States:

> We have yet to come up with a strong set of generalized norms, passed down through families, for savoring and sensibly consuming what our land and climate give us. We have, instead, a string of fad diets convulsing

our bookstores, one after another, at the scale of the national best seller. Nine out of ten nutritionists (unofficial survey) view this as evidence that we have lost our marbles. A more optimistic view might be this: these sets of mandates captivate us because we're looking hard for a food culture of our own. A profit-driven food industry has exploded and nutritionally bankrupted our caloric supply, and we long for a Food Leviticus to save us from the sinful roil of cheap fat and carbs. What the fad diets don't offer, though, is any sense of national and biological integrity. A food culture is not something that gets *sold* to people. It arises out of a place, a soil, a climate, a history, a temperament, a collective sense of belonging. Every set of fad-diet rules is essentially framed in the negative, dictating what you give up. [. . .] People hold to their food customs because of the *positives:* comfort, nourishment, heavenly aromas. (2007, 16–17)[16]

Do not get me wrong. I am not insinuating the vegetarianism is a fad diet. It is nothing of the sort. But I do suspect that its general appeal may have to do with the desire for a viable food culture just inasmuch as it has to do with rejecting the horrors of factory farming. Perhaps without being fully aware of it, vegetarians deeply desire a food culture that connects them with the land and with the multifarious people who inhabit it (ibid., 20).[17] Perhaps vegetarianism has become something of a stand-in for this.

Conceptualizing Totemism

Let me offer a possible bridging concept that can allow us to think about vegetarianism in terms that fit more comfortably with animism. Let us try out thinking about vegetarianism *totemically*. Totemism is a particular form of sociality expressed by animists, although it does have modern counterparts as well (Martin 2012).[18] Among animists, Deborah Bird Rose states that it constitutes "a non-random relationship between humans and particular non-humans" (1998). So totemism involves a spiritual, symbolic, and/or kinship connection between an individual or group and specifiable animals, plants, or earthly phenomena. Indeed, a totem is usually important because it bears a special relationship to an individual's or group's health and well-being (Cohan 2010, 76).[19]

According to Lupa, totems can take the form of archetypal beings akin to deities, individual spirits, or elements of our psyche. In each case, the totem "embodies the qualities of a given species of animal, plant, fungus, etc." (2014, 11).[20] Their physical counterparts are the living, breathing members of the community of life. Among their main functions is to act as intermediaries between humans and the other-than-human world (Lupa 2008, 147). Humans do well to listen to their totems, and the members of the other-than-human world rely on totems to learn more about us. For this reason, states Lupa, "When we work with totems, it isn't just beneficial to us" (2008, 148; see also Lupa 2014, 26). They provide a mechanism for us and for our totems' physical counterparts to hear and help one another.

Moreover, totemism connotes not just connectedness and interdependence with particular other-than-human beings but also the need to attend to specific responsibilities to them. "It organizes responsibilities for species along tracks that intersect, and thus builds a structure of regional systems of relationship and responsibility," Rose remarks (1998). In some contexts, it may not be possible to fulfill one's responsibilities without the cooperation of human kith and kin. In all animistic contexts, the bonds that tie individuals and groups to totems are interlaced with the land, or what Rose—with reference specifically to Australian Aboriginals—calls *country*. The focal point of law and life, one's country is small enough to accommodate face-to-face interactions between bands of people yet large enough to sustain their lives, autonomy, and structural equality. It is the place that gives and receives one's life and the lives of every member of one's community. So it is a multidimensional site of overlap among humans, other animals, plants, fungi, insects, Dream tracks (reflecting origin stories), soil, water, and air (Rose 1992 and Lupa 2012 and 2014).

The many kinds of people who populate a country take care of their own, Rose states. This also means that they know where their responsibilities end and others' responsibilities begin. But, continues Rose, "there are also responsibilities that are not linked to country, or that are specifically linked to a multiplicity of countries. [. . .] Thus, the very multiplicity of contexts through which the richness of life is structured ensures that no country/group is isolated, no boundary impermeable, no self-sufficiency total" (1998; see also Strehlow 1970). This creates a larger community composed by individuals and groups with intersecting responsibilities for the well-being of specific species or earthly phenomena. Those for whom rain is a totem, for example, are implicated in making rain for everyone. Those for whom the kangaroo is a totem depend on the rain totemists for precipitation while they devote their attention to kangaroos. And the well-being of the kangaroos benefits everybody, including the rain totemists.

These responsibilities also confer "powerful rights," Rose contends (ibid.), most notably the right to protect one's totem from harm. Protections, including restraint or prohibition of use, are socially enforceable among Australian Aboriginals. Seeking to protect one's totems can create tension and conflict, but there are ways to mitigate both. The overlap and intersection of rights claims require negotiation, Rose remarks. They also "ensure that power is located throughout the system." All are mutually dependent *and* mutually empowered. "In this system, living beings truly stand or fall together" (ibid.).

Thinking about Vegetarianism Totemically

The vegetarian impulse does not map onto totemism particularly fluidly. For one thing, among animists, totemic affiliations generally are inherited rather than chosen. For another, totemic affiliations are traditionally collective, not individual, manifestations. Among animists, they often are family, clan, or tribal denotations. But perhaps totemism nevertheless can provide a rough framework for understanding why a good many vegetarians maintain strongly held affinities to particular animals— namely, to the animals who most commonly become food for humans.

It is hardly unprecedented for people to discover their own totems, Lupa insists. We are all products of our culture, no matter how desperately we may desire to facilitate its demise. Indeed, facilitating its demise is only possible if we first acknowledge the position from which we are starting. The form of totemism that Lupa articulates takes full account of the individualistic and decentralized social conditions in which many of us live today. "What the totems tell me may not be what they tell you," she notes (ibid., 13). And surely what the totems tell the people of our culture can differ from what they tell the people of traditional cultures. It is hard to imagine that they do not have very different things to say to us—and very different things to say to their physical counterparts about us.

There is nothing wrong with informing ourselves about traditional totemic systems and learning how they operate. We surely need a new vision—a new story to enact (Quinn 1992)— if we are to abandon our old one. But we are bound to run into problems if we attempt simply to appropriate the totemic systems of indigenous cultures. We risk being complicit not only in the promotion of "plastic shamanism," or the commercialization of indigenous cultures, but also in embracing the false assumption that older totemic systems must be better than those that have emerged more recently. No totemic system is static. Traditions evolve among indigenous peoples, as they do for everyone, which leaves "a lot of room for growth and experimentation," Lupa asserts (2008, 18). Take, for example, the Nage of eastern Indonesia, who maintain a taboo against burning the wood and using the timber of the tamarind tree (from which they also take their tribal name). The sense of ethnic and sociopolitical identity of the Nage, barely a century old, is largely a function of colonialism. So is their deep affinity with the tamarind tree. Totemic taboos and relations can develop, and change, quite rapidly (Forth 2009).

We should keep this all in mind as we consider how to harness the vegetarian impulse to protect the animals who the people of our culture most regularly make our food. In consonance with the responsibilities that it engenders, totemism involves the delineation of specific behaviors toward the physical counterparts of one's totem. The comportment of the Nage toward the tamarind tree is a clear case in point. Correspondingly, for those with animal totems, this often involves blanket or contextual prohibitions against eating the totem's physical counterparts.[21]

Under specifiable conditions, perhaps vegetarians can be empowered to enact and enforce prospective taboos against killing and eating the physical counterparts of their totems (Latz 1995, 70, Harvey 2006, 100, and Cohan 2010, 76). This is necessary particularly when totemic populations are suffering—as are livestock in flesh factories today—or are endangered. Such an approach to restraining use, including setting aside places of refuge and protecting breeding sites, is likely to be good for the land and its inhabitants. If animals are raised for food where we live, at least in the manner that Fairlie advocates, this approach can even fit with what Lupa calls the *bioregional model* of totemism.[22] This model is predicated on the development and maintenance of an intimate relationship with our landbase and fellow inhabitants of it (2012, 3).

I suppose this entails that vegetarianism does not need to be defensible to be meaningful—at least in some limited sense. It would not do to overlook the ways in which the vegetarian impulse can serve the needs of the community of life. But we must take care not to let our sense of its meaningfulness carry us away.

CHAPTER 6

LOOSE ENDS

I mentioned in Chapter 3 that the animistic conceptual framework I offer is a loose mosaic that reflects points of intersection
and oft-repeated themes among numerous disparate animistic
traditions. It is not intended to capture every element, or even
every key element, of every animistic tradition. Nor do I suggest that I have presented the one right way to conceptualize
animism. How could I, since there is no one right way to be an
animist?

I also mentioned in Chapter 3 that developing a care-sensitive
relationship with who we make our food is no romantic or utopian notion. Even if we go on identifying ourselves by whether
plants or animals make up the last strand in the web of life
that leads to our mouths, we still can have a care-sensitive relationship with our food. The lives of the eaten would be better
off for it, and so would ours. Indeed, echoing my refrain from
Chapter 1, I hope I have made a compelling case that the world
would be made better—far better—if we could embrace that we
are full-fledged members of the community of life: constituents
of a closed-loop system from which we have borrowed, are now
using, and one day will return the fire of life that burns in us all.

I do not believe that my students would regard as shit what I
have offered here. They are fairly comfortable with the subject
of shit after all. But I would expect at least three complaints
from them at this point. Perhaps you have these complaints,

too. First, it has become so commonplace for proponents of vegetarianism to argue that their dietary proclivities are consonant with nonviolence and peaceful coexistence between humans and other-than-human beings that it has taken on an air of infallibility. So they are unlikely to be persuaded by my argument anyway, especially my suggestion that we are natural-born killers. Whether or not they are right to take the stand that they do is largely beside the point. What matters is that I risk alienating potential allies in my quest to get people to care more about who we eat and how we relate to who we eat.

Second, my critique of vegetarianism may provide omnivores with ammunition to defend refusing to change their dietary habits, despite the fact that I suggest making a wholesale change in how to view our relationship with who we make our food. If vegetarianism is morally indefensible and even ontologically illusory, I can imagine some omnivores concluding, then it turns out that how they eat *is* morally defensible—at least by comparison—even though I argue that omnivorism is ontologically illusory, too.

Third, I can hear my students asking, why not just be satisfied with the widespread abandonment of factory farming or, if we are really ambitious, conventional agriculture in all its forms? Even many dyed-in-the-wool omnivores would welcome one or both outcomes if for no other reason than that they would be able to enjoy considerably less-toxic food. So seeking outcomes like these is more practical than trying to get the people of our culture to embrace the primacy of our landbase, the personhood of fellow inhabitants of our landbase, our abiding membership in the community of life, and a food culture in which the distinctions between vegetarianism and omnivorism break down.

COMPETING WITHIN A
COOPERATIVE FRAMEWORK

I doubt that what I have to say in response to the first complaint will satisfy, or even appease, the vegetarians among you who remain unconvinced by my argument. I have no supererogatory obligation somehow to convince you to accept what I have to say, of course. You are as free to go your way as I am to go mine. But I can say unequivocally that I am just as concerned about creating conditions for peaceful coexistence between humans and other-than-human beings as you are. The problem we face is that our respective understandings of what this entails diverge in ways that may undermine what should be a fairly straightforward, if sometimes uneasy, alliance between us.

For the ontological vegan, for whom it is unjustifiable to use animals in any way, and the ontological vegetarian, for whom it is unjustifiable to eat animals, a basic precondition for peaceful coexistence with other-than-human animals requires that they not be killed or subject to unnecessary harm. Both parties may begrudgingly condone painlessly (or nearly painlessly) killing animals in a very limited set of cases, such as when one's survival is at stake. But in everyday situations, killing animals is off the table. This is not a matter for debate. And the prohibition against killing animals comes with a potential bonus, state some vegans and vegetarians, for it even can enhance our generaliz able sense of compassion or respect for fellow human beings. As a result, vegetarianism and veganism can provide salient bases for the promotion of a less violent, less militant world, according to their proponents.

This approach to seeking to deal peaceably with other-than-human animals is simple and straightforward. It is also more reflective of what comedian Stephen Colbert calls "truthiness" than of truth. "Truthiness," states Colbert, "is what you want the facts to be, as opposed to what the facts are. What feels like the right answer as opposed to what reality will support" (see Edwards 2008). We may want to be beings who do not have to kill—or at least to kill sentient beings—to survive. But in consonance with *kas-limaal*, this wish is a vestige of epistemic immaturity. To be sure, it is much easier to be a sentientist, or even an expansionary sentientist, than to engage in the hard work of uprooting the deep-seated axioms of our ecocidal culture. But nothing less is required, I submit, if we are to learn what living peaceably with other-than-human animals actually entails.

Recall if you will Fairlie's proposition that ours is a society that we should anticipate being in a state of energy descent, probably much sooner than we expect. Accommodating ourselves to this eventuality will require greater reliance on renewable energy, much better waste cycling, "and (consequently) a localized economy which is more integrated with natural processes" (2010, 100). If Fairlie is right, then we may not have the luxury to abstain from using, and perhaps even eating, other-than-human animals. Our landbase may dictate it. Without the ability to rely on a global economy to fulfill the demands of our dietary proclivities, we may have to reshape these proclivities if we are to attend to the health and well-being of our landbase. Simply put, our utter reliance on our landbase will become increasingly hard for us to ignore in the years and decades to come. So the conditions for peaceable living favored by vegetarians may not be suitable to surviving, let alone thriving.

I emphasize, though, that this does not mean we cannot deal peaceably with other-than-human animals. We most certainly can. Humans have done so for millions of years. It is the people of our culture who represent the exception, not the rule. I have offered one vision of what can work in this regard—for us, for others, and for the world—throughout *A Critique of the Moral Defense of Vegetarianism*. I give it one more quick shot here by taking a slightly different approach. Again, if the vegetarians among you remain unconvinced, so be it. At least I hope you understand where I am coming from.

The evolutionary development of every species that exists today is the product of two strategies that appear to contradict one another. But they do not. Every individual living being *competes* with other living beings for food, but this competition takes place within a wider *cooperative* framework.[1] The framework is life enhancing even if the process of competition between individuals involves taking life. Taking life, feeding and being fed upon, perpetuates the community of life. It shapes us all. In this respect, death can foster life. Go back and read the first section of Chapter 4 if you would like a reminder of the closed-loop system that constitutes the community of life.

Consider again the bacteria who populate our digestive system. Their evolutionary development has coincided with ours. Yes, this relationship is partially competitive. Our immune system has evolved in part to prevent them from eating us or monopolizing the food we ingest. This explains why antimicrobial acids are secreted in our stomachs. At the same time, though, the bacteria facilitate the production of digestive enzymes that break down our food for ready absorption in the lower digestive tract—which is where our "gut flora" reside. They feed on these "enzymic digestive products" (Keith 2009, 96) and, in the process, make them easier for us to digest.

Ruminants also benefit from the role microbes play in digestion. Consider how they draw sustenance from grasses, states Keith:

> The carbohydrate polymers that make up plant cell walls are indigestible to most animals and all mammals. Cellulose can only be broken down by microbial fermentation. The whole point of the ruminant's digestion is to keep food in the vast fermentative vat of its rumen so the bacteria have time to digest the cellulose. A cow regurgitates and rechews her food 500 times a day, for eight hours, approximating 25,000 chews. A cow is sacrificing the dietary protein in the grass, letting the microbes eat it instead. In the end, however, she trades in that poor quality plant protein for good quality microbial protein. This is what's happening inside a cow: she feeds grass to the bacteria, and then she eats them. (ibid., 96–97)

Think about how this process fits within the wider nutrient cycle of which we and ruminants are both part. We cannot photosynthesize. Neither can ruminants. But we and they can eat beings who can. We cannot digest cellulose. Ruminants can. And we are able to access the microbial protein that they create by eating them. And then the microbes that constitute the soil reclaim all of us to continue the whole process.

These are examples of what Quinn calls, variously, the "law of limited competition," the "peacekeeping law," and the "Law of Life." According to this law, "You may compete to the full extent of your capabilities, but you may not hunt down your competitors or destroy their food or deny them access to food. In other words, you may compete but you may not wage war" (1992, 129; see also Quinn 1996, 252). We humans came into existence following the law of limited competition. Animism is predicated on it. It specifies that some forms of killing, although not all, are consonant with peaceable living not just among competitors for food but also between predators and their prey. So in contrast to the commonplace assumption of many vegetarians, some forms of killing are indeed life enhancing. The war against the community of life that the people of our culture wage, of which factory farming is just one manifestation, represents a form of killing that vegetarians are absolutely right to condemn. This war is both morally abhorrent and evolutionarily unstable. The people of our culture are in the process of eliminating ourselves because we are systematically annihilating the community of life (Quinn 1996, 154). Factory farming is but one form of annihilation.

But we must not confuse waging war with peaceable, life-enhancing forms of killing. We have already seen one clear example of life-enhancing killing in our discussion of the reintroduction of wolves into Yellowstone National Park. And let us not ignore that we, too, can restore entire ecosystems by essentially *rewilding ourselves*. We can practice life-enhancing predation just as wolves do. We hereby can restore our broken agreements with our fellow members of the community of life and become members in good standing of it once again.

Hoagland and Dodson note that rewilding ourselves involves being self-interested (properly construed) rather than selfish, in the sense that we pay close attention to what facilitates our health and well-being: "Selfish behavior, pushed to the extreme, usually has unpleasant costs. A dominant animal engaging in too-frequent combat may sustain injuries. A parasite may kill its host and have nowhere to go. These self-defeating strategies generally get weeded out by evolution, so that in the long run most everyone tends to adopt some form of 'getting along'" (1995, 30). To hue one's own path unthinkingly at the expense of others is nothing less than gradual suicide (Justice 2011, 148). Wolves, for example, usually "take only the smallest, weakest, or most unhealthy of the prey species, leaving the fittest members to survive and reproduce," Hoagland and Dodson add. "This may be seen as being competitive at the individual level, cooperative at the group level. (Although we don't suggest that creatures generally think in terms of the group.)" (ibid.). Restrained predation of this sort is a recurring theme throughout the community of life.

Like most of the people of our culture, vegetarians may contend that other-than-human predators do not choose restraint. They may assume that the wolves' behavior is instead instinctual. Perhaps this is so, but perhaps not. Maybe this is yet another invidious example of anthropomorphism. What matters is that other-than-human predators *are* restrained. We became human be exhibiting restraint as well. We must kill to live, whether our prey is animal life or plant life. So we, the people of our culture, better relearn how to kill peaceably (and, somehow, force the hand of the noncompliant). We cannot choose to refrain from killing, as some vegetarians would have it, but we can—and must—choose to refrain from waging war.

A PYRRHIC VICTORY

Do I give fodder to omnivores seeking to defend the status quo? I would imagine that some of you worry that I do. Fox remarks that "Any challenge to meat eating can be expected to face considerable resistance, with corresponding attempts to undermine, discredit, marginalize, ridicule, silence, and intimidate those who have breached the assumed common ideology of human domination of nature" (1999, 30). He is surely right. I anticipate that some omnivores who are content with the status quo will happily appropriate from my argument what appeals to them while ignoring or simply dismissing what does not.

But they are hardly alone. Would you be surprised if some vegetarians do exactly what Fox describes with respect to my challenge to their alimentary proclivities? I would not be.[2] It is quite hard for just about anyone with deeply held convictions to do otherwise. This is a manifestation of commitment bias, according to which we are compelled to defend our beliefs simply because of the cumulative investment we already have made in them (Staw 1976). So our commitment in and of itself leads us to remain committed. Unconsciously or not, we seek out rationalizations that permit us to downplay countervailing evidence and neglect relevant information.

Commitment bias is a form of cognitive conservatism (which, to be clear, is not at all the same as political conservatism). In its most strident form, cognitive conservatism leads us only very reluctantly or begrudgingly to update our beliefs, if we do so at all (Sunstein 2000, 161ff., and Tetlock 2005, 126ff.). We also may incorporate updated beliefs into our wider belief network with as minimal a disruption as possible (James 1978, 35). I hope you believe me when I tell you that it took time even for me to accept my overarching conclusions. How could I expect anything less from you? So go ahead and rationalize if you will. Perhaps you will come around. Or perhaps you will give me reason to change my position. We should regard what I have to say here as part of an ongoing conversation after all. I certainly do not have the last word with respect to the complex and contentious issues I have addressed.[3]

Living in the Liminal

What about the third complaint—namely, that I should be satisfied with the widespread abandonment of factory farming, if not conventional agriculture in all its forms? For the record, I would be ecstatic were even the first of these to occur. Bringing down the animal-industrial complex would mark a major—indeed, an unprecedented—achievement in terms of ending the war the people of our culture are waging on the community of life. Frankly, though, I suspect that this may require upending our culture's core axioms anyway and hereby facilitating the collapse of our ecocidal culture.

Must an animistic culture similar to what I here describe necessarily replace our culture upon its collapse if we are to do well by our world? Of course not. Surely we have other options. I have outlined just one. But I admit that this concern, that I am asking for too much, troubles me deeply. Many of the people of our culture are very far away from having a genuinely care-sensitive relationship with who we make our food. This includes me. The structural barriers we face even to eradicating factory farming, let alone to facilitating the emergence of a food culture that works for people and for the planet, are prohibitively high. Our culture may not be sustainable, but at present it is proving to be quite resilient.

Perhaps even more daunting are the epistemic and experiential barriers to our embrace of an animistic culture or something like it. Indeed, it is these barriers that really give me pause when I reconsider whether what I offer here is utterly and completely quixotic after all. In my everyday life, I find it very difficult either to understand or to feel my embeddedness in the community of life. I am guessing that you do, too. The fact that I live in an urban setting certainly does not help. But my inabilities, and probably yours, run far deeper than this.

In one of his novels, Daniel Heath Justice writes of a character named Geth whose abiding sense of what it means to be an animist, or an adherent of the "Deep Green," is little more than "a tangled mix of romantic [. . .] tales, naïve fantasies, and superficial suppositions that had nothing to do with the long, often lonely work and painful transformations that the way of the Deep Green would demand of her" (2011, 44). This is due in part to the fact that Geth's culture had rather recently experienced a transformation whereby her kith and kin came to regard "the spirit people of the forests and mountains [as] enemies to be overcome rather than wisdom-bearers to be understood" (ibid., 34). Their gaze became transfixed by the supposedly "immortal heavens" (ibid., 24), and they "*forgot about their deep roots*" (ibid., 98).

I feel a lot like Geth right now, although my kith and kin lost their abiding connection to the Deep Green many generations ago. This is why the process of rewilding ourselves must involve more than a lifetime of effort. I thus offer these prescient words from Max Planck: "An important scientific innovation rarely makes its way by gradually winning over and converting its opponents. [. . .] What does happen is that its opponents die out and that the growing generation is familiarized with the idea from the beginning." The same surely can be said about cultural innovations like those I present in this book. If I cannot live in accordance with these innovations, can I at least do my best to help the next generation do so? Can I help them internalize their deep roots better than I can? To be honest, I do not know. But I damn well better try.

"We are living in the liminal," Natasha Alvarez proclaims (2014, xi), "standing at the threshold of changing our ways and complete annihilation. Which is it going to be?"

NOTES

CHAPTER 1

1. The context in which Jensen and McBay raise the question is grim. What, they ask, will be our legacy? What will we leave behind? We may do our best to live sustainably, but this is an impossible task so long as we inhabit and depend on a society and economy that are unsustainable—not to mention physically, socially, and psychologically toxic.
2. Lest you assume that notions like those articulated by Descartes and Bacon no longer have cultural currency, Jensen cites theoretical physicist Gerard Milburn, who states that "The aim of modern science is to reach an understanding of the world, not merely purely for aesthetic reasons, but that it may be ordered to our purpose" (2000, 20).

CHAPTER 2

1. Plant neurobiology bears no resemblance to the ideas proffered by Cleve Baxter (1968), who contends that plants react telekinetically to human thoughts and emotions. Nor do plant neurobiologists contend, like Peter Tomkins and Christopher Bird (1973), that plants have a "secret" psychical life. These sorts of propositions have not proven to be scientifically replicable.
2. To be a subject-of-a life, according to Regan, is to be a being who can experience pleasure and pain, perceive, remember, have a sense of the future, and be cognizant of one's psychophysical identity over time (1983, 243). This ostensibly marks a higher bar than sentience, but we will see that plants may well have many of these capacities anyway.
3. For Regan, we devalue subjects-of-a-life, and perhaps sentient beings as well, if we casually consume animals who may not quite meet comparable standards. We encourage the habit of viewing animals as mere means for our ends rather than ends in themselves. Singer's adherence to the precautionary principle is most evident with respect to his stance regarding mollusks.

He contends that "while one cannot with any confidence say that these creatures do feel pain, so can one equally have little confidence in saying that they do not feel pain. [. . .] Since it is so easy to avoid eating them, I now think it better to do so" (2004, 174).

4. "Lacking the ability to learn to distinguish between being touched by a harmless stimulus and by a potentially harmful stimulus," Deckers elaborates, "it [a plant] will respond in a similar fashion irrespective of the kind of stimulus that is provided. This lack of past experiences suggests that plants have more limited abilities to respond to external factors compared to animals and at least some other organisms. Since plants are less aware of their surroundings, it does not mean much to them to be controlled by external factors. Organisms who are capable of more complex modes of experiencing, however, can take in more information, and adapt themselves more rapidly and flexibly" (2009, 588).

5. The activation of nociceptors is not a conscious event per se, although the detection of noxious stimuli often is accompanied by conscious—typically painful—events.

6. Perhaps plants are not conscious in the sense of having something like subjective awareness of themselves experiencing the world, but they certainly can be awake and aware of the world around them. Interestingly, Marder notes that plants are much like us with respect to consciousness, at least in one sense: They sleep. "What largely determines vegetal activity above ground level is the plants' tending toward photosynthetic processing of sunlight. We would expect that in the absence of light, plants would be asleep. And they are! Anyone who has seen the time-lapse footage contrasting plant movements in the light and in the dark is struck by the difference between their purpose-oriented waking movements and the dreamy and somewhat chaotic vacillations at night" (2014, 182).

7. Auxins play a crucial role in coordinating plant growth and behavior, sometimes via long-distance signaling (Struik et al. 2008, 369). When the plant is wounded, auxins often induce cell differentiation and regeneration of vascular tissue.

8. Many plant scientists see such responses as reflexive. Phillips reports, though, that Trewavas sees them as flexible and adaptable. "Plasticity is foresight. Plants adjust their growth and development to maximize their fitness in a variable environment" (Phillips 2002, 40). This response is not unlike the contention by some philosophers that our free will is quite limited but that we have just as much of it as we need. It would be prohibitively time consuming if everything we do were to require a conscious

decision. Our biological "hardwiring" takes care of many life processes, which is an evolutionary advantage (Tancredi 2005). Perhaps this also is the case for plants.

9. "Brains come in handy for creatures that move around a lot," Pollan states, "but they're a disadvantage for ones that are rooted in place. Impressive as it is to us, self-consciousness is just another tool for living, good for some jobs, unhelpful for others" (2013, 92).

10. As Thomas Seeley and Royce Levien note, "It is not too much to say that a bee colony is capable of cognition in much the same way that a human being is. The colony gathers and continually updates diverse information about its surroundings, combines this with internal information about its internal state and makes decisions that reconcile its well-being with its environment" (1987, 39).

11. Fruitarianism may not be as free from harm as its proponents suggest anyway. In response to the claim by some fruitarians that fruit is the only freely given food, Lierre Keith remarks that "The point of fruit is not humans. The point is the seeds. The reason that the tree expends such tremendous resources accumulating fibers and sugars is to secure the best possible future for its offspring. And we take that offspring, in its swaddling of sweetness, and kill it" (2009, 14). This is because, unlike other animals, the people of our culture do not generally deposit seeds back in the earth. Even if we compost them, she adds, "time, heat, and bacteria kill them" (ibid., 13).

12. As Jensen comments, "Death is everywhere, and will seek me out no matter where I hide, now and again in the causing, and later in the receiving. This understanding came to me, oddly enough, when I was using the toilet. I realized that every time I defecate, I kill millions of bacteria. Every time I drink I swallow microorganisms, every time I scratch my head I kill tiny mites" (2000, 198).

CHAPTER 3

1. Plumwood has referred to herself as a "philosophical animist" (2013, 447). But it is not clear exactly what she means by this.

2. Plumwood provides examples like the following to demonstrate why the problem we should be concerned about is not the use of others per se but the "reductive treatment of the other as *no more than* something of use, as a means to an end" (2012, 82): "The circus performers who stand on one another's shoulders to reach the trapeze are not involved in any oppressively instrumental

practices. Neither is someone who collects animal droppings to improve a vegetable garden. In both cases the other is used, but is also seen as more than something to be used, and hence not treated instrumentally" (2004, 350).

3. Plumwood elaborates: "Although ontological vegans present themselves as extending our sympathies for our companion life forms on this planet, their rejection of sacred eating carries as its hidden underside another unstated project, that of narrowing and blunting our sympathies and sensibilities for the excluded class of living beings who are needed for our food" (2000, 302). I discuss sacred eating shortly.

4. I emphasize that attending with genuine concern to the needs and interests of plants just as we do to those of animals is not a "diversionary" tactic driven by "corpse eaters' fear of giving up the eating of animals," as Adams suggests (1994, 220, n. 60). Nor does giving plants their due somehow prevent us from maintaining different modes of solidarity with and different sorts of commitments to plants than we do to animals.

5. It is worth noting, in the words of Graham Harvey, that animism "is not an exercise in 'primitivism' because animism is far from primitive, nor is it about premodernity because animism does not serve as a precursor to modernity. Rather, animism is one of the many vitally present and contemporary other-than-modern ways of being human. Modernity rather than animism or any form of indigeneity is exceptional among ways of being human in this world" (2006, xxi).

6. Daniel Quinn states the following: "I can't imagine that anyone has undertaken either (1) to actually submit to aboriginal peoples worldwide any description or collection of descriptions of animism for them to rate as reflecting their own actual beliefs and perceptions or, (2) much better, to submit to them a carefully constructed test ('Do you agree completely, somewhat, or not at all with the following 215 statements?') that could actually serve as a scientific basis for saying what the commonality of worldview actually is among them. Until number 2 is done (or at least number 1), any definition of this invention called animism (including my own [. . .]) is ultimately nothing more than a fabrication that we outsiders find satisfactory" (personal communication 2015).

7. "There is nothing in these discourses that should be understood as implying (let alone asserting) that humans are the primary exemplars of personhood," Harvey declares. "[Irving] Hallowell's term 'other-than-human-person' celebrates two facts but does not confuse them: First, it arises from animist engagement

with a world that is full of persons, only some of whom are human; secondly, it arises from an animist acknowledgment that humans' most intimate relationships are had with other humans. Perhaps rock persons might speak of 'other-than-rock persons' while tree persons might speak of 'other-than-tree persons.' Such phrases, if unwieldy, are not intended to privilege any class of person but draw attention to degrees of relationality" (2006, xvii–xviii).

8. That *our* bonds of solidarity between different kinds of people may take varying forms does not necessitate that "the gods" care more for us and who we happen to privilege. "As the animist sees the world," states Quinn in direct contradistinction to Adams's considerations about plants, "everything that lives is sacred, the carrot no less than the cow. If there is any single doctrine that might win universal agreement among animists, I think it would be this, that the gods love everything that lives and have no favorites. If the gods have as much care for me as they do for a dandelion or a dragonfly, I'm perfectly content" (1994, 165).

9. For some animists, hunting may involve the consumption of other-than-human people to acquire their abilities. For others, argues Donna Haraway, one's prey must become "killable" (2007, 80), which may require what Eduardo Kohn calls their "desubjectivization" (2013, 114). This involves the transformation of an other-than-human person into a thing. This does not entail that the desubjectivized become mere means, though. They still qualify as more than food, which makes the act of hunting serious and often emotionally charged business. Moreover, animals do not start out being viewed as things, nor do they necessarily stay things. They may be resubjectivized under certain circumstances. As a result, Kohn states, "who counts as an *I* or a *you* and who becomes an *it* is relative and can shift. Who is predator and who is prey is contextually dependent" (ibid., 119).

10. Plumwood rejects what she refers to as "strong" panpsychism in favor of "weak" panpsychism. Proponents of the former advocate that mind in the form of something like a Cartesian res cogitans, or human-like cognition, permeates the cosmos. Proponents of the latter, including Plumwood (1993, 133), instead defend the proposition that mind-like qualities are widespread in the universe and not confined to the human sphere.

11. Were we compelled to hold on to the term, though, Plumwood suggests that "A more valid concept of anthropomorphism we might appeal to here would treat it as analogous, say, to certain ways of criticizing eurocentrism which would object to representing the non-European other in terms of a European norm" (2002, 59).

12. As Bird-David adds, "We do not personify other entities and then socialize with them but personify them *as, when,* and *because* we socialize with them" (1999, 77). Harvey remarks in turn that "Subjectivity itself is communal and continuously expressed in action, and found in the way people practice living towards other persons" (2006 113).

13. Deborah Bird Rose makes a similar claim about "ethical killing": "by which I mean killing that is responsive to the system of responsibilities and accountabilities" (2013, 143). Again, this is not made particularly easy by living at a distance from the processes that produce our food. As Jensen adds, "If your experience—far deeper than belief or perception—is that your food comes from the grocery store (and your water from the tap), from the economic system, from the social system we call civilization, it is to this you will pledge back your life. [. . .] If your experience—far deeper than belief or perception—is that food and water come from your landbase, or more broadly from the living earth, you will make and keep promises to your landbase in exchange for this food. You will honor and keep and participate in the fundamental predator/prey relationship. You will be responsible to the community that supplies you with food and water. You will defend this community to your very death" (2006, 696).

14. The scale of cities surely is unsustainable, but the structure—the way in which people dwell—may not be. The best way to feed locally in cities without outstripping the landbase is by expanding the size of the "landbase" by growing upward.

15. These three techniques have been used successfully around the world. With drip irrigation, plants root in troughs of lightweight, inert material, such as vermiculite, that can be used for years. Small tubes running from plant to plant drip nutrient-laden water at each stem's base, drastically reducing water waste in comparison with conventional irrigation techniques. During World War II, more than eight million pounds of vegetables were produced hydroponically on South Pacific islands for Allied forces there. With aeroponics, plants dangle in air that is infused with water vapor and nutrients, eliminating the need for soil, too.

16. When one asks if a process can be sufficiently scaled up, one tends to assume that we must be able to sustain our current population. But even if we challenge this assumption, it is important to assess the economic viability of large-scale vertical farming. Despommier notes that vertical farms can operate year-round at greater density than conventional farms. He estimates that

a thirty-story building covering one city block can produce roughly 2,400 acres of food per year. Growing can be further accelerated with 24-hour lighting. There have been considerable advances of late in lighting technology. Notably, LED lights have become increasingly efficient, providing higher output at lower costs (Usheroff 2013).

CHAPTER 4

1. As Jensen and McBay note, "It's estimated that in the U. S. alone, about 5.3 million gallons of embalming fluid are buried every year, enough to fill eight Olympic-sized swimming pools" (2009, 141). Along with formaldehyde, which itself is carcinogenic (and banned from use in Europe), embalming fluid may contain methanol, chloroform, chromate, toluene, methylene chloride, trichloroethylene, hexane, glutaraldehyde, and phenol. Add to this 30 million board feet of hardwood for caskets and 1.6 million tons of reinforced concrete for the burial vault. "Cremation is often promoted as a greener alternative," they continue, "and certainly it produces less groundwater pollution. But it also produces air pollution and uses the energy equivalent of forty-seven gallons of gasoline" (ibid., 143).
2. Ragan Sutterfield (2012) offers a contemporary reformulation of this idea.
3. This has political implications for Feuerbach. As Celimli-Inaltong remarks, "Because working-class people mainly live on potatoes, he [Feuerbach] states, which lack essential nutritional qualities, they lack the strength to overthrow the wealthy, thus lead a revolution" (2014, 1849). Improving the state of the working class, hence making it possible to reshape society in their image, first and foremost requires assuring that its constituents have access to more nutritious food (see also Hook 1994, 270).
4. Mathews's considerations bear the imprint of Hegel's dialectical logic, which I have no interest in appropriating. She also overlooks that the first two modalities actually share a deep allegiance to the cultural axioms identified by Quinn. But both of these matters are beside the point. For our purposes, I am most interested in her discussion of the third modality.
5. Interestingly, Mathews also notes that, like the animist, the postmaterialist has no need to impose an ecological ethic from without. To be postmaterialist is to engage with the world and with fellow inhabitants of it such that "we will not need any special department of environmentalism" (2006, 96).

6. The proper way to say this in order to highlight the transitivity of eating is as follows: We are who we eat. Who we eat is who we eat eats. So we are who who we eat eats. I avoid using this terminology given its obvious awkwardness.

7. Fox focuses wholly on the last of these attributions. He gives no attention to the underpinnings of tribal societies' forms of group identification and unification, which bear no clear connection to dominance.

8. Keith elaborates on this point: "The grazers need their daily cellulose, but the grass also needs the animals. It needs the manure, with its nitrogen, minerals, and bacteria; it needs the mechanical check of grazing activity; and it needs the resources stored in animal bodies and freed up by degraders when animals die" (2009, 8).

CHAPTER 5

1. Fairlie estimates that the feed-to-food ratio for pigs and chickens is 3:1 and 5:1, respectively. When we likewise include comestibles—wool, leather, manure, pet food, and so forth—he suggests that the overall feed-to-food ratio for meat production is roughly 1.4:1. Whether or not his statistics are entirely correct, he provides sufficient evidence to suggest that commonly cited statistics by defenders of vegetarianism are exaggerated. But significant inefficiencies certainly exist with industrialized meat production. How pigs are raised for food provides an interesting case in point. Monbiot (2010) notes the following: "Pigs [. . .] have been forbidden in many parts of the rich world from doing what they do best: converting waste into meat. Until the early 1990s, only 33 percent of compound pig feed in the UK consisted of grains fit for human consumption: the rest was made up of crop residues and food waste. Since then the proportion of sound grain in pig feed has doubled."

2. Fairlie suggests that it is more accurate to attribute 13.5 percent of global greenhouse gas emissions to meat production, which is still an awfully high figure. It drops to 12 percent if we separate out the production of soy on deforested Amazonian land, since it is used in many foods directly marketed to us rather than just as cattle feed. He concludes that "The only element of the FAO's 18 percent that would survive to a significant degree in a default livestock scenario would be methane emissions, mainly from ruminants' stomachs, which might remain equivalent to perhaps 4 percent of the world's greenhouse emissions" (Fairlie, Young, and Thomas 2008, 20). This number, too, is bound to be lower under conditions of energy descent like those discussed presently.

3. Clearly, a case can be made in favor of the health benefits of vegetarianism, which may diminish the ecological harm associated with reliance on a global food economy. Living a healthier life means having less need for pharmaceuticals. It also facilitates a longer life in which we are freer of the need for medical care, which is extremely resource intensive. But I think the argument in favor of the health benefits of vegetarianism is rather dubious. Yes, it is almost certainly healthier not to eat factory-farmed meat. But can the same be said for eating lean, organically raised animals—especially in moderation? A diet of this sort was presumably common among our ancestors, whose reproductive success is a necessary condition for our existence. My point is not that we should embrace a paleo diet. Attempting to do so is attended by its own problems (Zuk 2013). Rather, I am highlighting that if humans have gotten along for many millennia as eaters of the sort here described, I am not convinced that an appeal to the health benefits of vegetarianism can be used to push back against making use of a global food economy.

4. An alternative approach used to defend the ecological preferability of vegetarianism—and even more so veganism—is to measure the average number of animals killed to produce a comparable number of calories of different kinds of food. Mark Middleton (2009) provides a particularly detailed case in point. He concludes that the greatest amount of animal suffering and death can be prevented by embracing veganism. Middleton displays obvious sentientist sympathies. But I highlight instead that he voices no concern about vegans relying on a global food economy or conventional agriculture. So he fails to factor in habitat destruction, including the loss of waterways, watersheds, and even entire ecosystems, which is a common result of conventional farming.

5. As Jensen and McBay stress, "For an action to be sustainable, it can't merely parasitize off an unsustainable system. It must impede the system, and ultimately stop it from being unsustainable at all" (2009, 37). No action takes place in a vacuum. If the system on which we depend is not sustainable, then no action associated with it can be sustainable. Consider, notes Boyle (2011), that the vast majority of vegans fuel their cars with oil from oil companies that are responsible for an enormous number of animal deaths. He does not note this to denigrate veganism, mind you, but to highlight the difficulty all of us face in this culture with attempting to live by nonecocidal ideals.

6. Firor and Jacobsen expand on this point: "Each of these by itself can drive high fertility. Illiterate couples (and women in

particular) are less likely, even if they wish to limit childbearing, to know about family planning or its availability. Poor nutrition, inadequate sanitation, disease, and lack of health and other ser- vices to ameliorate these conditions all mean that more women face pregnancy with their physical resources compromised, more women die in childbirth, and more infants die before reaching age one. High infant mortality encourages high fertility; parents who have lost infants tend to have many children, even more than required to replace infants who die. Lack of schooling opportuni- ties means among other things that children cost less and provide more. They require no school fees and are able to work at home, and they grow up without the education that can break the cycle of high fertility, poor health, and poverty" (2002, 36).

7. Keith voices her agreement: "Two things work to stop overpop- ulation: ending poverty and ending patriarchy. People are poor because the rich are stealing from them. And most women have no control over how men use our bodies. If the major institu- tions around the globe would put their efforts behind initiatives like Iran's, there is still every hope that the world could turn toward both justice and sustainability" (2011, 228). In conso- nance, Richard Stearns states, "The single most significant thing that can be done to cure extreme poverty is this: protect, edu- cate, and nurture girls and women and provide them with equal rights and opportunities—educationally, economically, socially. [. . .] This one thing can do more to address extreme pov- erty than food, shelter, health care, economic development, or increased foreign assistance" (2010, 156–57).

8. See Quinn's "ABC's of ecology" (1996, 294ff.) for a more com- prehensive discussion of this phenomenon. As he states on his website, "When the food resources of a given species in the wild declines, its population declines for a number of reasons: more time must be spent searching for food, so there is less time for mating, females become less fertile, and less care is given the young, so that the population gradually declines" (http://www .ishmael.org/Interaction/QandA/Detail.CFM?Record=731). By comparison, mass starvation is not a common phenomenon.

9. It is true that the rate of human population increase is slowing. But the number of humans being added to our population is larger than at any time in human history, since the population base is larger than at any time in human history (Devall and Ses- sions 1985).

10. So we have not overshot human carrying capacity, as commenta- tors sometimes suggest (Jensen 2006, 123). This is not possible. We have overshot biomass population. But we are being carried

here and now—in increasing numbers for the time being—
because of our access to fossil fuels.

11. Zabel asserts that if biomass is the only energy source available to
us, human population will not grow very fast. Biomass population can grow at a slow rate so long as there is frontier land to
which populations can migrate. But if frontiers are fixed, growth
fluctuates around a given equilibrium.

12. In light of these considerations, you may wonder what the opti-
mum human population is. This depends in part on our standard
of living, states Alan Weisman. "At what level of material well-
being, and with what degree of distribution among the world's
people? With what technology, in what physical environments,
and with what kinds of governments? With what risk, robustness,
or stability [. . .]? And with what values?" (2013, 412).

13. Politics also can exacerbate—and has exacerbated—population
increase. In response to a study he conducted in Brazil in the
1960s, Herman Daly remarks that "The rich got richer while
the poor got children. An effective upper-class monopoly
on the means of limiting reproduction was added to the traditional monopoly on ownership of the means of production to
give an additional dimension of class domination" (1996, 119).
The same occurs today in areas otherwise dominated by the
poorest of the global poor. Quinn remarks, "If food doesn't
reach the starving millions, it is because they are too poor to
make it reach them. End poverty and there will be no starving
millions; everyone knows that there is enough food for them—if
they could afford to buy it. There are no starving *rich* people
anywhere in the world" (2010, 13).

14. This is perhaps why simply providing foreign aid to fight starvation often fails to alleviate the suffering of those who are most in
need. Michael Maren notes that aid instead ends up lining the
pockets of corrupt local officials and militias, agribusiness, the
media, and even aid organizations paid by the American government to distribute surplus food produced by subsidized American
farmers. "For television," asserts Maren, "the worst, most despair-
ing picture was the best famine and horror became a commodity. The worse it looked the better it sold. At the same time, the
possibility of American intervention increased the value of that
commodity" (1997, 212–13). In effect, Maren concludes, the
foreign aid industry has destroyed methods for dealing with
local food shortages that worked for centuries, if not millennia.
"Somalia in particular had a well-established system for dealing
with regular cycles of draught and famine. Farmers in the river
valleys built secure underground vaults where grain was stored

during fat years. When drought threatened the nomads, animals that might die anyway were exchanged for grains. Though nomads showed very little respect for farmers, they were aware that their lives might one day depend on these sedentary clans. They were therefore generous with the bounty of their herds when times were good. The result was a mutual insurance system and a truce of necessity across the land. Today, after huge infusions of international aid, Somalia and all its formerly self-sufficient neighbors are chronically hungry and dependent on foreign food. It becomes increasingly difficult for aid workers to ignore the compelling correlation between massive international food aid and increasing vulnerability to famine" (ibid., 21).

15. Bill McKibben (2007) offers as an example of ecological restoration the revitalization of the Adirondacks, where economic boom times are long past. The forest has rebounded, swallowing up numerous towns and encampments. Big cats and coyotes have been reintroduced and are thriving, even if their populations are small. This is compatible with what has come to be called *degrowth*, whose proponents insist that overconsumption can be combatted through economic contraction—through downscaling both production and consumption (D'Alisa et al. 2015). Ecological restoration also may be compatible with some rather unexpected practices. Consider the following from David Schmidtz and Elizabeth Willott: "We have no evidence of elephants ever being routinely hunted by any species other than humans. Just as in North America, where exterminating wolves and cougars caused deer and elk populations to explode, so too in parts of Africa where hunting by humans was stopped, elephant populations exploded. Without its keystone predator, any ecosystem is unstable. We can let natural processes control impala populations, but if we ask why we cannot likewise let natural processes control elephant populations, the answer is that, when it comes to elephants, hunting by humans *is* the natural process, or the closest thing to it. In Africa there is no such thing as humans simply 'letting nature be'" (2012, 466). The proliferation of elephants in these domains reduces biodiversity, Schmidtz and Willott contend. Moving herds often engenders acute social disruption among their members, and seeking to enlarge where herds roam may end up expanding the problem—if they continue to lack a predator. "Ultimately, something must play the role of that missing predator," Schmidtz and Willott conclude (ibid., 467).

16. Kingsolver elaborates: "Most people of my grandparents' generation had an intuitive sense of agricultural basics: when various fruits

and vegetables came into season, which ones keep over the winter, how to preserve the others. On what day autumn's first frost will likely fall on their county, and when to expect the last one in spring. [. . .] Most importantly: what animals and vegetables thrive in one's immediate region and how to live on those, with little else thrown into the mix beyond a bag of flour, a pinch of salt, and a handful of coffee. Few people of my generation, and approximately none of our children, could answer any of those questions, let alone all. This knowledge has vanished from our culture" (2007, 8–9). For the Baby Boom generation, education involved moving away from the dirt, Kingsolver argues. As far as I can tell, Millennials are returning to it, if in relatively modest numbers.

17. As Enrique Salmón states, "A key to any kind of restoration is to think about the traditional diets of communities. Once these diets are returned to the community it means that these people are living sustainably with their environment. The diet also does something else then, it's returning a form of nutrition that is unique to the specific place" (Martinez, Salmón, and Nelson 2008, 109–10). Keith remarks in turn that we should pose the following question to ourselves: "what grows where you live? Ask it, and you'll see. To answer, you will have to know the place you live. And if your food, your survival, is dependent upon the place that starts at your beating heart and extends as far as your legs can walk in a day, you will have to learn about rivers and forests, soil and rain" (2009, 56).

18. Lupa asserts that "totemism often started as a way to prevent people from marrying among their relatives; if a person had the same totem as you, you didn't get married. So in some cultures, the totem plays the role of embodied surname. This all weaves into the original meaning of the word *totem*, which derives from an Ojibwe root word concerning relationship with something or someone" (2012, 9; see also Levi-Strauss 1971). Historically, the anthropological study of totemism focuses on how people classify themselves—as individuals and much more often as groups—with reference to the community of life (Radcliffe-Brown 1952, 116, and Borneman 1988). This assumes a human-centered understanding of the world—a projection of intra- and inter-group relations onto the community of life (Foucault 1973)—that is foreign to many animistic traditions.

19. Taking their cue mainly from Claude Levi-Strauss, numerous anthropologists contend that totemism is a useless or empty conceptual category. But the concept nevertheless "obstinately refuses to 'lie down,'" states Roy Willis (1990, 5). The study of totemism is perhaps even undergoing a modest revival among

scholars. But we should acknowledge Levi-Strauss's concerns. Most notably, he argues that the study of totemism cannot yield a completely general, unitary, or systematic principle of social or conceptual order. It also has not proven useful as a category for cross-cultural analysis. Whether these are shortcomings is a matter that requires far more consideration than I can give here. I note, though, that animism yields no general or unitary principle either but is still quite useful. Moreover, Gregory Forth declares with respect to totemism, "To the extent that it can have comparative value, the term is best employed as an 'odd-job' word denoting a variety of relations between social groups (and especially social subgroups or segments) and natural kinds of phenomena" (2009, 264).

20. Archetypal beings "embody all the qualities of a given animal—Bear as opposed to a bear spirit," Lupa contends (2012, 8).

21. This sort of comportment toward the physical counterpart of totems is not universal. Some peoples eat physical counterparts to maintain their relationship with their totem (Saunders 1995). As Lupa states, "if we look at the case of the Lakota, the bison was (and still is) one of the most sacred animals in their totemic system precisely because they killed and ate it. The bison represented the life of this culture" (2008, 60).

22. Lupa compares the bioregional model to the correspondence and archetypal models. The first "uses directions, elements, and other patterns and correspondences to create a sort of totemic cosmology" (2012, 3). The second "plugs into the human psyche, both individually and by way of the collective unconscious, the part of our consciousness shared with other humans, inherited from our ancestors" (ibid., 3–4).

CHAPTER 6

1. Lynn Margulis and Dorion Sagan (1995) assert that a key mechanism driving the evolution of early life on Earth was two or more species initially working together and then permanently joining together as a single, more complex life form.

2. I would not say that I count on receiving challenges of this sort as I write this final chapter. I do hope for them, though. I do not mean to suggest that I look forward to being ridiculed or marginalized. My point, rather, is that it is difficult for me to imagine that my work will be of much help in challenging our cultural axioms if some—and perhaps many—people are not prepared to storm my proverbial barricades with pitchforks in hand.

3. Perhaps some people are uncomfortable with the fact that I do not come down cleanly in favor of either vegetarianism or omnivorism. Recall the black-and-white approach that Adams takes with respect to the moral standing of plants, apparently in order to leave no conceptual wiggle room for defenders of the animal-industrial complex to exploit. I have little to say in response to this concern that I have not said already. In one respect, we must be willing to live with nuance. In another, though, my argument is not nuanced at all, since I reject the entire conceptual framework according to which the distinction between vegetarianism and omnivorism is morally or ontologically viable.

BIBLIOGRAPHY

Abram, D., *The Spell of the Sensuous: Perception and Language in a More-than-Human World* (New York: Vintage, 1996).

Adams, C. J., *The Sexual Politics of Meat: A Feminist-Vegetarian Critical Theory* (New York: Continuum, 1990).

Adams, C. J., "The Feminist Traffic in Animals," in *Ecofeminism: Women, Animals, Nature,* ed. G. Gaard (Philadelphia: Temple University Press, 1993), pp. 195–218.

Adams, C. J., *Neither Man nor Beast: Feminism and the Defense of Animals* (New York: Continuum, 1994).

Adams, C. J., "Feeding on Grace: Institutional Violence, Christianity, and Vegetarianism," in *Religious Vegetarianism: From Hesiod to the Dalai Lama,* ed. K. S. Walters and L. Portmess (Albany, NY: SUNY Press, 2001), pp. 148–68.

Adamson, R., "First Nations Survival and the Future of the Earth," in *Original Instructions: Indigenous Teachings for a Sustainable Future,* ed. M. K. Nelson (Rochester, VT: Bear, 2008), pp. 27–35.

Alpi, A., et al., "Plant Neurobiology: No Brain, No Gain?" *Trends in Plant Science* 12 (2007): 135–36.

Altieri, M. A., *Agroecology* (Boulder, CO: Westview, 1995).

Alvarez, N., *Liminal: A Novella* (Ephrata, PA: Black and Green, 2014).

Amato, P. R., and S. A. Partridge, *The New Vegetarians* (New York: Plenum, 1989).

Appel, H. M., and R. B. Cocroft, "Plants Respond to Leaf Vibrations Caused by Insect Herbivore Chewing," *Oecologia* 175 (2014): 1257–66.

Ariely, D., *Predictably Irrational: The Hidden Forces that Shape Our Decisions* (New York: Harper Perennial, 2010).

Arnold, P. P., and A. G. Gold, *Sacred Landscapes and Cultural Politics: Planting a Tree* (Aldershot, UK: Ashgate, 2001).

Baluška, F., "Recent Surprising Similarities between Plant Cells and Neurons," *Plant Signaling Behavior* 5 (2010): 87–89.

Baluška, F., S. Lev-Yadun, and S. Mancuso, "Swarm Intelligence in Plants," *Trends in Ecology and Evolution* 25 (2010): 682–83.

Baluška, F., and S. Mancuso, "Plant Neurobiology: From Stimulus Perception to Adaptive Behavior of Plants, via Integrated Chemical and Electrical Signaling," *Plant Signaling and Behavior* 4 (2009): 475–76.

Baluška, F., D. Volkmann, A. Hlavacka, S. Mancuso, and P. W. Barlow, "Neurobiological View of Plants and Their Body Plan," in *Communication in Plants: Neuronal Aspects of Plant Life*, ed. F. Baluška, S. Mancuso, and D. Volkmann (Berlin: Springer-Verlag, 2006), pp. 19–23.

Baluška, F., D. Volkmann, and D. Menzel, "Plant Synapses: Actin-Based Domains for Cell-to-Cell Communication," *Trends in Plant Science* 10 (2005): 106–11.

Barlow, P. W., "Reflections on 'Plant Neurobiology,'" *BioSystems* 92 (2008): 132–47.

Baxter, C., "Evidence of a Primary Perception in Plant Life," *International Journal of Parapsychology* 10(4) (1968): 329–48.

Bechta, R. L., and W. J. Ripple, "River Channel Dynamics following Extirpation of Wolves in Northwestern Yellowstone National Park," *Earth Surface Processes and Landforms* 31(12) (2006): 1525–39.

Bell, D., *Daughters of the Dreaming* (North Melbourne, AU: Spinifex, 2001).

Benhim, J. K. A., "Agriculture and Deforestation in the Tropics: A Critical and Empirical Review," *AMBIO* 35(1) (2006): 9–16.

Berndt, R. M., and C. H. Berndt, *The Speaking Land: Myth and Story in Aboriginal Australia* (New York: Penguin, 1989).

Bernstein, M., "Contractualism and Animals," *Philosophical Studies* 86 (1997): 49–72.

Bhalla, U. S., and R. Iyengar, "Emergent Properties of Networks of Biological Signaling Pathways," *Science* 283 (1999): 381–87.

Biao, X., W. Xiaorong, D. Zhuhong, and Y. Yaping, "Critical Impact Assessment of Organic Agriculture," *Journal of Agricultural and Environmental Ethics* 16 (2003): 297–311.

Biedrzycki, B., "Kin Recognition in Plants: A Mysterious Behavior Unsolved," *Journal of Experimental Botany* 61 (2010): 4123–28.

Bird-David, N., "Animism Revisited: Personhood, Environment, and Relational Epistemology," *Current Anthropology* 40 (1999): S67–91.

Bird-David, N., "Animistic Epistemology: Why Do Some Hunter-Gatherers Not Depict Animals?" *Ethnos* 71(1) (2006): 33–50.

Bird-David, N., and D. Naveh, "Relational Epistemology, Immediacy, and Conservation: Or, What Do the Nayaka Try to Conserve?" *Journal of the Study of Religion, Nature, and Culture* 2(1) (2008): 55–73.

Black, M. B., "Ojibwa Power Belief System," in *The Anthropology of Power*, ed. R. D. Fogelson and R. N. Adams (New York: Academic Press, 1977), pp. 141–51.

Bonan, G., "Forests and Climate Change: Forcing, Feedbacks, and the Climate Benefits of Forests," *Science* 320(5882) (2008): 1444–49.

Borneman, J., "Race, Ethnicity, Species, Breed: Totemism and Horse-Breed Classification in America," *Contemporary Studies in Society and History* 30(1) (1988): 25–51.

Boyle, M., *The Moneyless Man: A Year of Freeconomic Living* (Oxford, UK: Oneworld, 2010).

Boyle, M., "Have Simon Fairlie & George Monbiot Got It Wrong about Meat Eating?" *Permaculture* (February 16, 2011), http://www.permaculture.co.uk/articles/have-simon-fairlie-george-monbiot-got-it-wrong-about-meat-eating.

Bradshaw, C. J. A., and B. W. Brook, "Human Population Reduction Is Not a Quick Fix for Environmental Problems," *Proceedings of the National Academy of Sciences* 111(46) (2014): 16610–15.

Bradshaw, C. J. A., M. A. Hindell, N. J. Best, K. L. Phillips, G. Wilson, and P. D. Nichols, "You Are What You Eat: Describing the Foraging Ecology of Southern Elephant Seals (*Mirounga leonine*) Using Blubber Fatty Acids," *Proceedings of the Royal Society of London B* 270(1521) (2003): 1283–92.

Brandt, K., and J. P. Mølgaard, "Organic Agriculture: Does It Enhance or Reduce the Nutritional Value of Plant Food?" *Journal of the Science of Food and Agriculture* 81 (2001): 924–31.

Braunstein, M. M., *Radical Vegetarianism*, rev. ed. (Quaker Hill, CT: Panacea, 1993).

Brenner, E. D., R. Stahlberg, S. Mancuso, F. Baluška, and E. Von Volkenburgh, "Response to Alpi et al.: Plant Neurobiology: The Gain Is More than the Name," *Trends in Plant Science* 12 (2007): 285–86.

Brenner, E. D., R. Stahlberg, S. Mancuso, J. Vivanco, F. Baluška, and E. Von Volkenburgh, "Plant Neurobiology: An Integrated View of Plant Signaling," *Trends in Plant Science* 11 (2006): 413–19.

Brillat-Savarin, J. A., *The Physiology of Taste: Or Meditations on Transcendental Gastronomy,* trans. M. F. K. Fisher (New York: Vintage, 2011).

Brown, D., *An Interpretation of the Social Theories and Novels of Daniel Quinn: How Can We Create a Sustainable Society?* (Lewiston, NY: Edwin Mellen, 2009).

Brown, L. R., *Eco-Economy: Building an Economy for the Earth* (New York: W. W. Norton, 2001).

Buhaug, H., and H. Urdal, "An Urbanization Bomb? Population Growth and Social Disorder in Cities," *Global Environmental Change* 23(1) (2013): 1–10.

Buhner, S. H., *The Lost Language of Plants: The Ecological Importance of Plant Medicines to Life on Earth* (White River Junction, VT: Chelsea Green, 2002).

Callaway, R. M., "Positive Interactions among Plants," *The Botanical Review* 61(4) (1995): 306–49.

Callaway, R. M., "The Detection of Neighbors by Plants," *Trends in Ecology and Evolution* 17 (2002): 104–5.

Callaway, R. M., S. C. Pennings, and C. L. Richards, "Phenotypic Plasticity and Interactions among Plants," *Ecology* 84 (2003): 1115–28.

Callicott, J. B., *In Defense of the Land Ethic: Essays in Environmental Philosophy* (Albany, NY: SUNY Press, 1989).

Calvo Garzón, P., and F. Keijzer, "Cognition in Plants," in *Plant–Environment Interactions,* ed. F. Baluška (Berlin: Springer-Verlag, 2009), pp. 247–66.

Cardinale, B. J., J. P. Wright, M. W. Cadotte, I. T. Carroll, A. Hector, D. S. Srivastava, M. Loreau, and J. J. Weis, "Impacts of Plant Diversity on Biomass Production Increase through Time Because of Species Complementarity," *Proceedings of the National Academy of Sciences* 104(46) (2007): 18123–28.

Celimli-Inaltong, I., "You Are What You Eat," in *Encyclopedia of Food and Agricultural Ethics,* ed. P. B. Thompson and D. M. Kaplan (New York: Springer, 2014), pp. 1845–50.

Chamowitz, D., *What a Plant Knows* (New York: Scientific American / Farrar, Straus and Giroux, 2012).

Chernela, J., "Piercing Distinctions: Making and Remaking the Social Contract in the North-West Amazon," in *Beyond the Visible and the Material,* ed. L. M. Rival and N. L. Whitehead (New York: Oxford University Press, 2001), pp. 177–95.

Ciszak, M., D. Comparini, B. Mazzolai, F. Baluška, F. T. Arecchi, T. Vicsek, and S. Mancuso, "Swarm Behavior in Plant Roots," *PLoS ONE* 7 (2012): e29759.

Cockrall-King, J., *Food and the City: Urban Agriculture and the New Food Revolution* (Amherst, NY: Prometheus, 2012).

Cohan, J. A., *The Primitive Mind and Modern Man* (Sharjah, UAE: Bentham Science, 2010).

Cohen, J. E., "Human Population: The Next Half Century," *Science* 302(5648) (2003): 1172–75.

Cole, T., "The White-Savior Industrial Complex," *The Atlantic* (March 21, 2012), http://www.theatlantic.com/international/archive/2012/03/the-white-savior-industrial-complex/254843/.

Commoner, B., *The Closing Circle: Nature, Man, and Technology* (New York: Random House, 1971).

Commoner, B., "How Poverty Breeds Overpopulation (And Not the Other Way Around)," *Ramparts* 13(10) (1975): 21–24, 59.

Connor, D. J., "Organic Agriculture Cannot Feed the World," *Field Crops Research* 106 (2008): 187–90.

Crowder, D. W., T. D. Northfield, M. R. Strand, and W. E. Snyder, "Organic Agriculture Promotes Evenness and Natural Pest Control," *Nature* 466 (2010): 109–12.

Cummins, R., and B. Lilliston, *Genetically Modified Food* (New York: Marlow, 2000).

Curnutt, J., "A New Argument for Vegetarianism," in *Disputed Moral Issues*, ed. M. Timmons (New York: Oxford University Press, 2011), pp. 362–72.

Curtin, D., "Contextual Moral Vegetarianism," in *Food for Thought: The Debate over Eating Meat*, ed. S. F. Sapontzis (Amherst, NY: Prometheus, 2004), pp. 272–83.

D'Alisa, G., F. Demaria, and G. Kallis, eds., *Degrowth: A Vocabulary for a New Era* (New York: Routledge, 2015).

Daly, H., *Beyond Growth: The Economic of Sustainable Development* (Boston: Beacon, 1996).

Darwin, C., *The Power of Movement in Plants* (London: J. Murray, 1880).

Davis, S. L., "The Least Harm Principle May Require that Humans Consume a Diet Containing Large Herbivores, Not a Vegan Diet," *Journal of Agricultural and Environmental Ethics* 16(4) (2003): 387–94.

Deckers, J., "Vegetarianism, Sentimental or Ethical?" *Journal of Agricultural and Environmental Ethics* 22 (2009): 573–97.

DeGrazia, D., *Taking Animals Seriously* (New York: Cambridge University Press, 1996).

de Kroon, H., and M. J. Hutchings, "Morphological Plasticity in Clonal Plants: The Foraging Concept Reconsidered," *Journal of Ecology* 83 (1995): 143–52.

Descartes, R., *Discourse on Method,* in *The Philosophical Writings of Descartes,* vol. 1, trans. J. Cottingham, R. Stoothoff, and D. Murdoch (New York: Cambridge University Press, 1985).

de Schutter, O., *Report Submitted by the Special Rapporteur on the Right to Food* (New York: United Nations, 2010).

Descola, P., *Beyond Nature and Culture,* trans. J. Lloyd (Chicago: University of Chicago Press, 2013).

Despommier, D., "The Rise of Vertical Farms," *Scientific American* 301(5) (2009): 80–87.

Despommier, D., *The Vertical Farm: Feeding the World in the 21st Century* (New York: St. Martin's Press, 2010).

Despommier, D., "Farming up the City: The Rise of Urban Vertical Farms," *Trends in Biotechnology* 31(7) (2013): 388–89.

Detwiler, F., "All My Relatives: Persons in Oglala Religion," *Religion* 22 (1992): 235–46.

Devall, B., and G. Sessions, *Deep Ecology: Living as if Nature Mattered* (Salt Lake City: Peregrine Smith, 1985).

Devine, P. E., "The Moral Basis of Vegetarianism," *Philosophy* 53 (1978): 481–505.

Dicke, M. J., and J. Bruin, "Chemical Information Transfer between Damaged and Undamaged Plants," *Biochemical Systematics and Ecology* 29 (2001): 979–1113.

Duram, L. A., *Good Growing* (Lincoln: University of Nebraska Press, 2005).

Durning, A. B., and H. B. Brough, *Taking Stock: Animal Farming and the Environment* (Washington, DC: Worldwatch Institute, 1991).

Dziubinska, H., "Ways of Signal Transmission and the Physiological Role of Electrical Potentials in Plants," *Acta Societatis Boticorum Poloniae* 72 (2003): 309–18.

Edwards, J., "Truthiness and Consequences in the Public Use of Reason: Useful Lies, a Noble Lie, and a Supposed Right to Lie," *Veritas* 53(1) (2008): 73–91.

Ehrenreich, B., "Maid to Order," in *Global Woman: Nannies, Maids, and Sex Workers in the New Economy,* ed. B. Ehrenreich and A. R. Hochschild (New York: Henry Holt, 2002), pp. 85–103.

Ehrlich, P., *The Population Bomb: Population Control or Race to Oblivion?* (New York: Ballantine, 1968).

Engel, M., Jr., "The Immorality of Eating Meat," in *The Moral Life*, ed. L. P. Pojman (New York: Oxford University Press, 2000), pp. 856–90.

Ensminger, A. H., *Foods and Nutrition Encyclopedia*, 2nd ed. (Boca Raton, FL: CRC Press, 1993).

Fairlie, S., *Meat: A Benign Extravagance* (White River Junction, VT: Chelsea Green, 2010).

Fairlie, S., R. Young, and P. Thomas, "Getting to the Meat of the Matter," *The Ecologist* 38(8) (2008): 14–23.

Farmer, E. E., and C. A. Ryan, "Interplant Communication: Airborne Methyl Jasmonate Induces Synthesis of Proteinase Inhibitors in Plant Leaves," *Proceedings of the National Academy of Sciences of the USA* 87 (1990): 7713–16.

Farrington, B., *The Philosophy of Francis Bacon* (Liverpool: Liverpool University Press, 1964).

Fiddes, N., *Meat: A Natural Symbol* (New York: Routledge, 1991).

Firn, R., "Plant Intelligence: An Alternative Point of View," *Annals of Botany* 93 (2004): 345–51.

Firor, J., and J. Jacobsen, *The Crowded Greenhouse: Population, Climate Change, and Creating a Sustainable World* (New Haven, CT: Yale University Press, 2002).

Fischer, B., and J. Tronto, "Toward a Feminist Theory of Caring," in *Circles of Care: Work and Identity in Women's Lives*, ed. E. Abel and M. Nelson (Albany, NY: SUNY Press, 1990), pp. 36–54.

Food and Agriculture Organization of the United Nations, *Livestock's Long Shadow: Environmental Issues and Options* (New York: Author, 2006).

Forbes, J. D., *Columbus and Other Cannibals* (New York: Seven Stories Press, 2008).

Foreman, D., *Rewilding North America: A Vision for Conservation in the 21st Century* (Washington, DC: Island Press, 2004).

Forth, G., "Tree Totems and the Tamarind People: Implications of Clan Plant Taboos in Central Flores," *Oceania* 79(3) (2009): 263–79.

Foucault, M., *The Order of Things: An Archeology of the Human Sciences* (New York: Vintage, 1973).

Fox, M. A., *Deep Vegetarianism* (Philadelphia: Temple University Press, 1999).

Fox, M. A., "Vegetarianism and Planetary Health," *Ethics and the Environment* 5 (2000): 163–74.

Fox, M. A., "Why We Should Be Vegetarians," *International Journal of Applied Philosophy* 20 (2006): 295–310.

Francione, G. L., *Introduction to Animal Rights: Your Child or Your Dog?* (Philadelphia: Temple University Press, 2000).

Friedman, T. L., *Hot, Flat, and Crowded: Why We Need a Green Revolution—and How It Can Renew America* (New York: Picador, 2009).

Fromm, J., and S. Lautner, "Electrical Signals and Their Physiological Significance in Plants," *Plant, Cell and Environment* 30 (2007): 249–57.

Gagliano, M., et al., "Experience Teaches Plants to Learn Faster and Forget Slower in Environments Where It Matters," *Oecologia* 175 (2014): 63–72.

Geber, M. A., M. A. Watson, and H. de Kroon, "Organ Preformation, Development, and Resource Allocation in Perennials," in *Plant Resource Allocation*, ed. F. A. Bazzaz and J. Grace (San Diego: Academic, 1997), pp. 113–43.

Gersani, M., et al., "Tragedy of the Commons as a Result of Root Competition," *Ecology* 89 (2001): 660–69.

Gilman, S. L., *Diets and Dieting: A Cultural Encyclopedia* (New York: Routledge, 2008).

Gollner, A. L., *The Fruit Hunters* (New York: Scribner, 2008).

Grewal, S. S., and P. S. Grewal, "Can Cities Become Self-Reliant in Food?" *Cities* 29(1) (2012): 1–11.

Grime, J. P., and J. M. L. Mackey, "The Role of Plasticity in Resource Capture by Plants," *Evolutionary Ecology* 16 (2002): 299–307.

Gruntman, M., and A. Novoplansky, "Physiologically Mediated Self/ Nonself Discrimination in Roots," *PNAS* 101 (2004): 2863–67.

Gruzalski, B., "Why It's Wrong to Eat Animals Raised and Slaughtered for Food," in *Food for Thought: The Debate over Meat Eating*, ed. S. F. Sapontzis (Amherst, NY: Prometheus, 2004), pp. 124–37.

Guillebaud, J., and P. Hayes, "Population Growth and Climate Change: Universal Access to Family Planning Should Be the Priority," *Free Inquiry* 29(3) (2009): 34–35.

Gunn Allen, P., "The Psychological Landscape of *Ceremony*," *American Indian Quarterly* 5(1) (1979): 7–12.

Guss, D. M., *To Weave and Sing: Art, Symbol, and Narrative in the South American Rain Forest* (Berkeley: University of California Press, 1989).

Haddad, N. M., et al., "Habitat Fragmentation and Its Lasting Impact on Earth's Ecosystems," *Science Advances* 1(2) (2015): e1500052.

Hall, M., *Plants as Persons* (New York: Columbia University Press, 2011).

Hall, M., "Talk among the Trees: Animist Plant Ontologies and Ethics," in *The Handbook of Contemporary Animism*, ed. G. Harvey (Bristol, CT: Acumen, 2013), pp. 385–94.

Hallowell, A. I., "Ojibwa Ontology, Behavior, and World View," in *Culture in History*, ed. S. Diamond (New York: Columbia University Press, 1960), pp. 19–52.

Haraway, D., *Modest_Witness@Second_Millennium.FemaleMan_Meets _OncoMouse: Feminism and Technoscience* (New York: Routledge, 1997).

Haraway, D., *When Species Meet* (Minneapolis: University of Minnesota Press, 2007).

Hardin, G., "Living on a Lifeboat," *Bioscience* 24 (1974): 561–68.

Hartmann, T., *The Last Hours of Humanity* (Cardiff, CA: Waterfront Digital Press, 2013).

Harvey, G., *Animism: Respecting the Living World* (New York: Columbia University Press, 2006).

Harvey, G., "Introduction," in *The Handbook of Contemporary Animism*, ed. Graham Harvey (Bristol, CT: Acumen, 2013), pp. 1 12.

Hawken, P., *The Ecology of Commerce: A Declaration of Sustainability*, rev. ed. (New York: Harper Business, 2010).

Hayden, D., *The Power of Place* (Cambridge, MA: MIT Press, 1995).

Henare, M., "Tapu, Mana, Mauri, Hau, Wairua: A Maori Philosophy of Vitalism and Cosmos," in *Indigenous Traditions and Ecology*, ed. J. A. Grim (Cambridge, MA: Harvard University Press, 2001), pp. 197–221.

Hill, J. L., *The Case for Vegetarianism* (Lanham, MD: Rowman and Littlefield, 1996).

Hoagland, M., and B. Dodson, *The Way Life Works* (New York: Times Books, 1995).

Hoff, C., "Immoral and Moral Uses of Animals," *New England Journal of Medicine* 302 (1980): 115–18.

Hogan, L., *Dwellings: A Spiritual History of the Living World* (New York: W. W. Norton, 1995)

Hogan, L., "We Call It Tradition," in *The Handbook of Contemporary Animism*, ed. Graham Harvey (Bristol, CT: Acumen, 2013), pp. 17–26.

Holden, C., and J. MacDonald, *Nutrition and Child Health* (Amsterdam: Bailliere Tindall, 2000).

Hook, S., *From Hegel to Marx: Studies in the Intellectual Development of Karl Marx* (New York: Columbia University Press, 1994).

Hopfenberg, R., and D. Pimenthel, "Human Population Numbers as a Function of Food Supply," *Environment, Development, and Sustainability* 3 (2001): 1–15.

Hursthouse, R., "Virtue Ethics and the Treatment of Animals," in *The Oxford Handbook of Animal Ethics,* ed. T. L. Beauchamp and R. G. Frey (New York: Oxford University Press, 2011), pp. 119–43.

Hutchings, M. J., and H. de Kroon, "Foraging in Plants: The Role of Morphological Plasticity in Resource Acquisition," *Advances in Ecological Research* 25 (1994): 159–238.

Izaguiree, M. M., et al., "Remote Sensing of Future Competitors: Impacts on Plant Defences," *PNAS* 103 (2006): 7170–74.

Jackson, R. B., and M. M. Caldwell, "Integrating Resource Heterogeneity and Plant Plasticity: Modeling Nitrate and Phosphate Uptake in a Patchy Soil Environment," *Journal of Ecology* 84 (1996): 891–903.

Jackson, W., "Nature as a Measure of Sustainable Agriculture," in *Environmental Ethics: The Big Questions,* ed. D. R. Keller (Malden, MA: Wiley-Blackwell, 2010), pp. 476–81.

Jamail, D., "'The Great Dying' Redux? Shocking Parallels between Ancient Mass Extinction and Climate Change," *Salon* (December 17, 2013), http://www.salon.com/2013/12/17/the_great_dying _redux_shocking_parallels_between_ancient_mass_extinction_and _climate_change_partner/.

James, W., *Pragmatism* and *The Meaning of Truth* (Cambridge, MA: Harvard University Press, 1978).

James, W., *A Pluralistic Universe* (Lincoln: University of Nebraska Press, 1996).

Jensen, D., *A Language Older than Words* (White River Junction, VT: Chelsea Green, 2000).

Jensen, D., *Endgame,* vol. 1 and 2 (New York: Seven Stories Press, 2006).

Jensen, D., and A. McBay, *What We Leave Behind* (New York: Seven Stories Press, 2009).

Justice, D. H., *The Way of Thorn and Thunder: The Kynship Chronicles* (Albuquerque: University of New Mexico Press, 2011).

Karban, R., and K. Shiojiri, "Self-Recognition Affects Plant Communication and Defense," *Ecology Letters* 12 (2009): 502–6.

Kaufman, S., *Reinventing the Sacred: A New View of Science, Reason, and Religion* (New York: Basic Books, 2008).

Keith, L., *The Vegetarian Myth: Food, Justice, and Sustainability* (Oakland, CA: Flashpoint Press, 2009).

Keith, L., "Other Plans," in *Deep Green Resistance,* by A. McBay, L. Keith, and D. Jensen (New York: Seven Stories Press, 2011), pp. 193–237.

Keller, D. R., and F. B. Golley, *The Philosophy of Ecology: From Science to Synthesis* (Athens: University of Georgia Press, 2000).

Keller, E. F., *A Feeling for the Organism: The Life and Work of Barbara McClintock* (San Francisco: W. H. Freeman, 1983).

Kelly, C. L., "Plant Foraging: A Marginal Value Model and Coiling Response in Cuscuta Subinclusa," *Ecology* 71 (1990): 1916–25.

Kheel, M., "License to Kill: An Ecofeminist Critique of Hunters' Discourse," in *Animals and Women: Feminist Theoretical Explorations,* ed. C. J. Adams and J. Donovan (Durham, NC: Duke University Press, 1995), pp. 85–125.

Kheel, M., "Vegetarianism and Ecofeminism: Toppling Patriarchy with a Fork," in *Food for Thought: The Debate over Eating Meat,* ed. S. F. Sapontzis (Amherst, NY: Prometheus, 2004), pp. 327–41.

Kingsolver, B., *Animal, Vegetable, Miracle: A Year of Food Life* (New York: HarperCollins, 2007).

Kohák, E., "Varieties of Ecological Experience," *Environmental Ethics* 19 (1997): 153–71.

Köhler, A., "On Apes and Men: Baka and Bantu Attitudes to Wildlife and the Making of Eco-Goodies and Baddies," *Conservation and Society* 3(2) (2005): 407–35.

Kohn, E., *How Forests Think: Toward an Anthropology beyond the Human* (Berkeley: University of California Press, 2013).

Kohn, M. J., "You Are What You Eat," *Science* 283(5400) (1999): 335–36.

Kover, T. R., "Flesh, Death, and Tofu: Hunters, Vegetarians, and Carnal Knowledge," in *Hunting: In Search of the Wild Life,* ed. N. Kowalsky (Malden, MA: Wiley-Blackwell, 2010), pp. 171–83.

Ladner, P., *The Urban Food Revolution: Changing the Way We Feed Cities* (Gabriola Island, BC: New Society, 2011).

Latz, P., *Bushfires and Bushtucker: Aboriginal Plant Use in Central Australia* (Alice Springs, AU: IAD Press, 1995).

Laws, R., "Native Americans and Vegetarianism," *Vegetarian Journal* (September 1994), http://www.ivu.org/history/native_americans.html.

Leopold, A., *A Sand County Almanac* (New York: Oxford University Press, 1949).

Levi-Strauss, C., *Totemism* (Boston: Beacon, 1971).

Liang, D., and J. Silverman, "'You Are What You Eat': Diet Modifies Cuticular Hydrocarbons and Nestmate Recognition in the Argentine Ant, *Linepithema humile*," *Naturwissenschaften* 87(9) (2000): 412–16.

Liebman, M., and E. Dyck, "Crop Rotation and Intercropping Strategies for Weed Management," *Ecological Applications* 3 (1993): 92–122.

Light, A., "Climate Ethics for Climate Action," in *Environmental Ethics: What Really Matters, What Really Works*, ed. D. Schmidtz and E. Willott (New York: Oxford University Press, 2012), pp. 557–66.

Llorente, R., "The Moral Framework of Peter Singer's *Animal Liberation*: An Alternative to Utilitarianism," *Ethical Perspectives* 16 (2009): 61–80.

Lovewisdom, J., *The Ascensional Science of Spiritualizing Fruitarian Dietetics* (San Francisco: Paradisian, 2005).

Lupa, *DIY Totemism: Your Personal Guide to Animal Totems* (Stafford, UK: Megalithica, 2008).

Lupa, *New Paths to Animal Totems: Three Alternative Approaches to Creating Your Own Totemism* (Woodbury, MN: Llewellyn, 2012).

Lupa, *Plant and Fungus Totems: Connect with Spirits of Field, Forest, and Garden* (Woodbury, MN: Llewellyn, 2014).

Lyon, P., "The Biogenic Approach to Cognition," *Cognitive Processing* 7(1) (2006): 11–29.

Maina, G. G., J. S. Brown, and M. Gersani, "Intra-Plant versus Inter-Plant Competition in Beans: Avoidance Resource Matching or Tragedy of the Commons," *Plant Ecology* 160 (2002): 235–47.

Marder, M., "Is Plant Liberation on the Menu?" *New York Times* (May 8, 2012a), http://opinionator.blogs.nytimes.com/2012/05/08/is-plant-liberation-on-the-menu/?_php=true&_type=blogs&_r=0.

Marder, M., "Plant Intentionality and the Phenomenological Framework of Plant Intelligence," *Plant Signaling and Behavior* 7 (2012b): 1365–72.

Marder, M., *Plant-Thinking* (New York: Columbia University Press, 2013).

Marder, M., *The Philosopher's Plant: An Intellectual Herbarium* (New York: Columbia University Press, 2014).

Marder, M., and G. Francione, "Michael Marder and Gary Francione Debate Plant Ethics," *Columbia University Press Blog* (March 6, 2013), http://www.cupblog.org/?p=9605.

Maren, M., *The Road to Hell: The Ravaging Effects of Foreign Aid and International Charity* (New York: Free Press, 1997).

Margulis, L., and D. Sagan, *What Is Life?* (New York: Simon and Schuster, 1995).

Martin, C., "On Totems of Science and Capitalism: Or, Why We Are All 'Religious,'" *Implicit Religion* 15(1) (2012): 25–35.

Martinez, D., E. Salmón, and M. K. Nelson, "Restoring Indigenous History and Culture to Nature," in *Original Instructions: Indigenous Teachings for a Sustainable Future*, ed. M. K. Nelson (Rochester, VT: Bear, 2008), pp. 88–115.

Martinez, J. L., "Environmental Pollution by Antibiotics and by Antibiotic Resistant Determinants," *Environmental Pollution* 157(11) (2009): 2893–902.

Marx, K., "Economic and Philosophical Manuscripts," trans. L. D. Easton and K. H. Guddat, in *Writings of the Young Marx on Philosophy and Society*, ed. L. D. Easton and K. H. Guddat (Indianapolis: Hackett, 1997), pp. 283–337.

Matheny, G., "Least Harm: A Defense of Vegetarianism from Steven Davis's Omnivorous Proposal," *Journal of Agricultural and Environmental Ethics* 16 (2003): 505–11.

Mathews, F., *For Love of Matter: Contemporary Panpsychism* (Albany, NY: SUNY Press, 2003).

Mathews, F., "Beyond Modernity and Tradition: A Third Way for Development," *Ethics and Environment* 11(2) (2006): 85–113.

McCabe, J., *Sunfood Living* (Lakeside, CA: Sunfood, 2007).

McDonough, W., and M. Braungart, *Cradle to Cradle: Remaking the Way We Make Things* (New York: North Point Press, 2002).

McGaa, E., *Mother Earth Spirituality* (San Francisco: HarperCollins, 1990).

McIntyre, B. D., et al., *International Assessment of Agricultural Knowledge, Science and Technology for Development* (Washington, DC: Island, 2009).

McKibben, B., *Hope, Human and Wild* (Minneapolis, MN: Milkweed, 2007).

McKibben, B., *Eaarth: Making a Life on a Tough New Planet* (New York: Times Books, 2010).

Middleton, M., "The Number of Animals Killed to Produce One Million Calories in Eight Food Categories," *Animal Visuals* (October 17, 2009), http://www.animalvisuals.org/projects/data/1mc.

Midgley, M., *Animals and Why They Matter* (Athens, GA: University of Georgia Press, 1983).

Midgley, M., *Utopias, Dolphins, and Computers: Problems in Philosophical Plumbing* (New York: Routledge, 1996).

Modesto, R., *Not for Innocent Ears* (Angelus Oaks, CA: Sweetwater, 1980).

Mohawk, J., "The Art of Thriving in Place," in *Original Instructions: Indigenous Teachings for a Sustainable Future*, ed. M. K. Nelson (Rochester, VT: Bear, 2008), pp. 126–36.

Molinier, J., G. Ries, C. Zipfel, and B. Hohn, "Transgeneration Memory of Stress in Plants," *Nature* 442 (2006): 1046–49.

Mollison, B., *Permaculture: A Designer's Manual* (Tyalgum, AU: Tagari, 1988)

Monbiot, G., "I Was Wrong about Veganism. Let Them Eat Meat—But Farm It Properly," *The Guardian* (September 6, 2010), http://www.theguardian.com/commentisfree/2010/sep/06/meat-production-veganism-deforestation?CMP=share_btn_link.

Monbiot, G., *Feral: Rewilding the Land, the Sea, and Human Life* (Chicago: University of Chicago Press, 2014).

Mowat, F., *People of the Deer* (New York: Little, Brown, 1952).

Nagel, T., *Mind and Cosmos* (New York: Oxford University Press, 2012).

Naveh, D., and N. Bird-David, "Animism, Conservation, and Immediacy," in *The Handbook of Contemporary Animism*, ed. G. Harvey (Bristol, CT: Acumen, 2013), pp. 27–37.

Neal, A. L., S. Ahmad, R. Gordon-Weeks, and J. Ton, "Benzoxazinoids in Root Exudates of Maize Attract *Pseudomonas putida* to the Rhizosphere," *PLoS ONE* 7(4) (2012): e35498.

Neal, A. L., and J. Ton, "Systemic Defense Priming by *Pseudomonas putida* KT2440 in Maize Depends on Benzoxazinoid Exudation from the Roots," *Plant Signaling and Behavior* 8(1) (2013): 120–24.

Nelson, J., "8,500 Gallons of Water for 1 Pound of Beef," *Vegsource* (2004), http://www.vegsource.com/articles2/water_stockholm.htm.

Nelson, M. K., "Lighting the Sun of Our Future—How These Teachings Can Provide Illumination," in *Original Instructions:*

Indigenous Teachings for a Sustainable Future, ed. M. K. Nelson (Rochester, VT: Bear, 2008a), pp. 1–19.

Nelson, M. K., "Re-Indigenizing Our Bodies and Minds through Native Foods," in *Original Instructions: Indigenous Teachings for a Sustainable Future,* ed. M. K. Nelson (Rochester, VT: Bear, 2008b), pp. 180–95.

Noble, D., *The Music of Life* (New York: Oxford University Press, 2006).

Noddings, N., "Comment on Donovan's 'Animal Rights and Feminist Theory,'" *Signs* 16 (1991): 418–22.

Nonhebel, S., "On Resource Use in Food Production Systems: The Value of Livestock as 'Rest-Stream Upgrading System,'" *Ecological Economics* 48 (2004): 221–30.

Ortega y Gasset, J., *Meditations on Hunting,* trans. Howard B. Wescott (New York: Charles Scribner's Sons, 1985).

Ouma, G., and P. Jeruto, "Sustainable Horticultural Crop Production through Intercropping: The Case of Fruits and Vegetable Crops: A Review," *Agriculture and Biology Journal of North America* 1 (2010): 1098–105.

Paré, P. W., and J. H. Tumlinson, "Plant Volatiles as a Defense against Insect Herbivores," *Plant Physiology* 121 (1999): 325–32.

Phillips, H., "Not Just a Pretty Face: They May Be Green, but Plants Aren't Stupid. Helen Phillips Picks Their Brains," *New Scientist* 175 (July 27, 2002): 40.

Piaget, J., *The Child's Conception of the World* (London: Kegan Paul, 1929).

Piaget, J., *The Construction of Reality in the Child* (New York: Basic Books, 1954).

Picasso, V. D., E. C. Brummer, M. Liebman, P. M. Dixon, and B. J. Wilsey, "Crop Species Diversity Affects Productivity and Weed Suppression in Perennial Polycultures under Two Management Strategies," *Crop Science* 48(1) (2008): 331–42.

Pimentel, D., D. Cerasale, R. C. Stanley, R. Perlman, E. M. Newman, L. C. Brent, A. Mullan, and D. Tai-I Chang, "Annual vs. Perennial Grain Production," *Agriculture, Ecosystems and Environment* 161 (2012) 1–9.

Pluhar, E., "Who Can Be Morally Obligated to Be a Vegetarian?" *Journal of Agricultural and Environmental Ethics* 5 (1992): 189–215.

Pluhar, E., "The Right to Not Be Eaten," in *Food for Thought: The Debate over Eating Meat,* ed. S. F. Sapontzis (Amherst, NY: Prometheus, 2004), pp. 92–107.

Plumwood, V., *Feminism and the Mastery of Nature* (New York: Routledge, 1993).

Plumwood, V., "Integrating Ethical Frameworks for Animals, Humans, and Nature: A Critical Feminist Eco-Socialist Analysis," *Ethics and the Environment* 5(2) (2000): 285–322.

Plumwood, V., *Environmental Culture: The Ecological Crisis of Reason* (New York: Routledge, 2002).

Plumwood, V., "Animals and Ecology: Toward a Better Integration," in *Food for Thought: The Debate over Eating Meat,* ed. S. F. Sapontzis (Amherst, NY: Prometheus, 2004), pp. 344–58.

Plumwood, V., "Shadow Places and the Politics of Dwelling," *Australian Humanities Review* 44 (2008): 139–50.

Plumwood, V., *The Eye of the Crocodile,* ed. L. Shannon (Canberra, AU: ANU Press, 2012).

Plumwood, V., "Nature in the Active Voice," in *The Handbook of Contemporary Animism,* ed. G. Harvey (Bristol, CT: Acumen, 2013), pp. 441–53.

Pollan, M., *The Omnivore's Dilemma: A Natural History in Four Meals* (New York: Penguin, 2006).

Pollan, M., "The Intelligent Plant," *The New Yorker* 89 (December 23, 2013): 92.

Porphyry, *On Abstinence from Animal Food,* trans. T. Taylor (Watchung, NJ: Albert Saifer, 1989).

Postel, S., *Pillar of Sand* (New York: W. W. Norton, 1999).

Preece, R., *Animals and Nature: Cultural Myths, Cultural Realities* (Vancouver: UBC Press, 1999).

Preece, R., "Ask Your Brother for Forgiveness: Animal Respect in Native American Traditions," in *Food for Thought: The Debate over Eating Meat,* ed. S. F. Sapontzis (Amherst, NY: Prometheus, 2004), pp. 236–46.

Pretchel, M., *Long Life, Honey in the Heart* (Berkeley, CA: North Atlantic, 2004).

Pringle, L., *The Animal Rights Controversy* (San Diego: Harcourt Brace, 1989).

Quinn, D., *Ishmael* (New York: Bantam, 1992).

Quinn, D., *Providence: The Story of a Fifty-Year Vision Quest* (New York: Bantam, 1994).

Quinn, D., *The Story of B* (New York: Bantam, 1996).

Quinn, D., *Beyond Civilization: Humanity's Next Great Adventure* (New York: Three Rivers, 1999).

Quinn, D., *The Book of the Damned* (Houston: New Tribal Ventures, 2001a).

Quinn, D., *The Man Who Grew Young* (New York: Context, 2001b).

Quinn, D., *If They Give You Lined Paper, Write Sideways* (Hanover, NH: Steerforth, 2007).

Quinn, D., "The Danger of Human Exceptionalism," in *Moral Ground: Ethical Action for a Planet in Peril,* ed. K. D. Moore and M. P. Nelson (San Antonio, TX: Trinity University Press, 2010), pp. 9–14.

Rachels, J., "Do Animals Have a Right to Life?" in *Ethics and Animals,* ed. H. B. Miller and W. H. Williams (Clifton, NJ: Humana, 1983), 275–84.

Rachels, J., "The Basic Argument for Vegetarianism," in *Food for Thought: The Debate over Eating Meat,* ed. S. F. Sapontzis (Amherst, NY: Prometheus, 2004), pp. 70–80.

Rachels, S., "Vegetarianism," in *The Oxford Handbook of Animal Ethics,* ed. T. L. Beauchamp and R. G. Frey (New York: Oxford University Press, 2011), pp. 877–905.

Radcliffe-Brown, A. R., *Structure and Function in Primitive Society* (New York: Free Press, 1952).

Reedy, A., ed., *Ngo Korero a Pita Kapiti: The Teachings of Pita Kapiti* (Christchurch, NZ: University of Canterbury Press, 1997).

Regan, T., *The Case for Animal Rights* (Berkeley, CA: University of California Press, 1983).

Regan, T., "The Rights of Humans and Other Animals," *Ethics and Behavior* 7 (1997): 103–11.

Regan, T., *The Case for Animal Rights,* 2nd ed. (Berkeley, CA: University of California Press, 2004).

Regenstein, L., "Animal Rights, Endangered Species, and Human Survival," in *In Defense of Animals,* ed. P. Singer (New York: Basil Blackwell, 1985), pp. 118–32.

Rifkin, J., *Beyond Beef* (New York: Plume, 1992).

Ripple, W. J., and R. L. Beschta, "Trophic Cascades in Yellowstone: The First 15 Years after Wolf Reintroduction," *Biological Conservation* 145(1) (2012): 205–13.

Robbins, J., "2,500 Gallons All Wet?" *EarthSave* (n.d.), http://www.earthsave.org/environment/water.htm.

Roberts, J. A., K. A. Elliott, and Z. H. Gonzales-Carranza, "Abscission, Dehiscence, and Other Cell Separation Processes," *Annual Review of Plant Biology* 53 (2002): 131–58.

Rolston, H., III, *Environmental Ethics: Duties to and Values in the Natural World* (Philadelphia: Temple University Press, 1988).

Rolston, H., III, "Treating Animals Naturally?" *Between the Species* 5 (1989): 131–32.

Rose, D., *Dingo Makes Us Happy* (New York: Cambridge University Press, 1992).

Rose, D. B., "Totemism, Regions, and Co-Management in Aboriginal Australia," paper presented at the Conference of the International Association for the Study of Common Property, 1998.

Rose, D. B., "Death and Grief in a World of Kin," in *The Handbook of Contemporary Animism*, ed. G. Harvey (Bristol, CT: Acumen, 2013), pp. 137–47.

Rowlands, M., *Animals Like Us* (New York: Verso Press, 2002).

Salmón, E., "Kincentric Ecology: Indigenous Perceptions of the Human-Nature Relationship," *Ecological Applications* 10(5) (2000): 1327–32.

Samour, P., *Handbook of Pediatric Nutrition* (Sudbury, MA: Jones and Bartlett, 2005).

Sanchez, C. L., "Animal, Vegetable, Mineral: The Sacred Connection," in *Ecofeminism and the Sacred*, ed. C. J. Adams (New York: Continuum, 1999), pp. 207–28.

Sapontzis, S. F., *Morals, Reasons, and Animals* (Philadelphia: Temple University Press, 1987).

Saunders, N. J., *Living Wisdom: Animal Spirits* (Boston: Little, Brown, 1995).

Savory, A., *The Grazing Revolution: A Radical Plan to Save the Earth* (New York: TED Books, 2013).

Schleifer, H., "Images of Death and Life: Food Animal Production and the Vegetarian Option," in *In Defense of Animals*, ed. P. Singer (New York: Basil Blackwell, 1985), pp. 62–73.

Schmidtz, D., and E. Willott, "Reinventing the Commons: An African Case Study," in *Environmental Ethics: What Really Matters, What Really Works*, ed. D. Schmidtz and E. Willott (New York: Oxford University Press, 2012), pp. 458–71.

Schull, J., "Are Species Intelligent?" *Behavioral and Brain Sciences* 13 (1990): 63–108.

Schwartz, J., *Cows Save the Planet: And Other Improbable Ways for Restoring Soil to Heal the Earth* (White River Junction, VT: Chelsea Green, 2013).

Seeley, T. D., and R. A. Levien, "A Colony of Mind: The Beehive as Thinking Machine," *The Sciences* 27 (1987): 39–42.

Sen, A., *Poverty and Famines* (New York: Oxford University Press, 1981).

Sen, A., "Population: Delusion and Reality," *The New York Review of Books* 41(15) (September 22, 1994), http://www.nybooks.com/articles/archives/1994/sep/22/population-delusion-and-reality.

Seufert, V., N. Ramankutty, and J. A. Foley, "Comparing the Yields of Organic and Conventional Agriculture," *Nature* 485 (May 10, 2012): 229–32.

Sexton, R., "Abscission," in *Handbook of Plant and Crop Physiology*, ed. M. Pesserakli (New York: Dekker, 1995), pp. 497–525.

Shafer-Landau, R., "Vegetarianism, Causation, and Ethical Theory," *Public Affairs Quarterly* 8 (1994): 85–100.

Shiva, V., "Biotechnology and Genetically Modified Organisms: Solutions or New Problems?" *Impulso* 15 (2004): 35–43.

Shiva, V., "Myths about Industrial Agriculture," *Al Jazeera* (September 23, 2012), http://www.aljazeera.com/indepth/opinion/2012/09/2012998389284146.html.

Shoumatoff, A., "This Rare, White Bear May Be the Key to Saving a Canadian Rainforest," *Smithsonian Magazine* (September 2015), http://www.smithsonianmag.com/science-nature/rare-white-bear-key saving-canadian-rainforest-180956330/?no-ist.

Singer, P., "Killing Humans and Killing Animals," *Inquiry* 22 (1979): 145–56.

Singer, P., *Animal Liberation* (New York: HarperCollins, 2002).

Singer, P., "Animal Liberation: Vegetarianism as Protest," in *Food for Thought: The Debate over Eating Meat*, ed. S. F. Sapontzis (Amherst, NY: Prometheus, 2004), pp. 108–17.

Singer, P., *The Expanding Circle: Ethics, Evolution, and Moral Progress* (Princeton, NJ: Princeton University Press, 2011).

Smil, V., *Enriching the Earth: Fritz Haber, Carl Bosch, and the Transformation of World Food Production* (Cambridge, MA: MIT Press, 2001).

Smith, A., "For All Those Who Were Indian in a Former Life," in *Ecofeminism and the Sacred*, ed. C. J. Adams (New York: Continuum, 1993), pp. 168–71.

Spencer, C., *The Heretic's Feast: A History of Vegetarianism* (Hanover, NH: University Press of New England, 1995).

Staw, B. M., "Knee-Deep in the Big Muddy: A Study of Escalating Commitment to a Chosen Course of Action," *Organizational Behavior and Human Performance* 16(1) (1976): 27–44.

Stearns, R., *The Hole in Our Gospel: What Does God Expect of Us? The Answer That Changed My Life and Might Just Change the World* (New York: Thomas Nelson, 2010).

Steiner, G., *Animals and the Moral Community* (New York: Columbia University Press, 2008).

Stengers, I., "Reclaiming Animism," *e-flux* 36 (2012), http://www.e-flux.com/journal/reclaiming-animism.

Stenhouse, D., *The Evolution of Intelligence* (New York: Harper and Row, 1974).

Sterba, J. P., *Contemporary Social and Political Philosophy* (Belmont, CA: Wadsworth, 1995).

Stiller, J. W., "Plastid Endosymbiosis, Genome Evolution, and the Origin of Green Plants," *Trends in Plant Science* 12 (2007): 391–96.

Strehlow, T., "Geography and Totemic Landscape in Central Australia: A Functional Study," in *Australian Aboriginal Anthropology: Modern Studies in Social Anthropology of the Australian Aborigines*, ed. R. M. Berndt (Canberra, AU: Australian Institute of Aboriginal Studies, 1970), pp. 124–26.

Struik, P. C., X. Yin, and H. Meinke, "Plant Neurobiology and Green Plant Intelligence: Science, Metaphor, and Nonsense," *Journal of the Science of Food and Agriculture* 88 (2008): 363–70.

Sumner, L. W., "Animal Welfare and Animal Rights," *Journal of Medicine and Philosophy* 13 (1988): 159–75.

Sung, S., and R. M. Amasino, "Molecular Genetic Study of the Memory of Winter," *Journal of Experimental Botany* 57 (2000): 3369–77.

Sung, S., and R. M. Amasino, "Vernalisation and Epigenetics: How Plants Remember Winter," *Current Opinion in Plant Biology* 7 (2004): 4–10.

Sunstein, C. R., *Behavioral Law and Economics* (New York: Cambridge University Press, 2000).

Susiarjo, M., and M. S. Bartolomei, "You Are What You Eat, but What about Your DNA?" *Science* 345(6198) (2014): 733–34.

Sutterfield, R., "You Are Who You Eat," *Sustainable Traditions* (August 17, 2012), http://sustainabletraditions.com/2012/08/you-are-who-you-eat/.

Talisse, R. B., *Democracy and Moral Conflict* (New York: Cambridge University Press, 2009).

Tancredi, L., *Hardwired Behavior: What Neuroscience Reveals about Morality* (New York: Cambridge University Press, 2005).

Tawhai, T. P., "Maori Religion," in *The Study of Religion: Traditional and New Religion,* ed. S. Sutherland and P. Clarke (New York: Routledge, 1988), pp. 96–105.

Taylor, P. W., *Respect for Nature* (Princeton, NJ: Princeton University Press, 1986).

Tetlock, P. E., *Expert Political Judgment: How Good Is It? How Can We Know?* (Princeton, NJ: Princeton University Press, 2005).

Thaler, J. S., "Jasmonate-Inducible Plant Defences Cause Increased Parasitism of Herbivores," *Nature* 399 (June 17, 1999): 686–88.

Thomas, D., "A Study on the Mineral Depletion of the Foods Available to Us as a Nation over the Period 1940 to 1991," *Nutrition and Health* 17 (2003): 85–115.

Thompson, M. V., and N. M. Holbrook, "Scaling Phloem Transport: Information Transmission," *Plant, Cell and Environment* 27 (2004): 509–19.

Tomkins, P., and C. Bird, *The Secret Life of Plants: A Fascinating Account of the Physical, Emotion, and Spiritual Relations between Plants and Man* (New York: Harper and Row, 1973).

Trewavas, A., "Urban Myths of Organic Farming," *Nature* 410 (2001): 409–10.

Trewavas, A., "Plant Intelligence," *Naturwissenschaften* 92 (2005): 401–13.

Trewavas, A., "Response to Alpi et al.: Plant Neurobiology—All Metaphors Have Value," *Trends in Plant Science* 12 (2007): 231–33.

Trewavas, A., "What Is Plant Behaviour?" *Plant, Cell, and Environment* 32 (2009): 606–16.

Trewavas, A., "Plants Are Intelligent Too," *EMBO Reports* 13 (2012): 772–73.

Turner, B. S., *The Body and Society* (London: Sage, 1996).

Tverberg, G., "The Long-Term Tie between Energy Supply, Population, and the Economy," *Our Finite World* (August 29, 2012), http://ourfiniteworld.com/2012/08/29/the-long-term-tie-between-energy-supply-population-and-the-economy/.

Usheroff, M., "Sky's the Limit for Vertical Farming: Crops Take Less Space, Less Water," *Orange County Register* (December 7, 2013), https://global-factiva-com.ezproxy2.library.drexel.edu/ga/default.aspx.

Vanacek, J., "Nutrigenomics: You Really Are What You Eat," *Forbes* (October 9, 2012), http://www.forbes.com/sites/sap/2012/10/09/nutrigenomics-the-study-of-you-are-what-you-eat/.

VanDeVeer, D., "Interspecific Justice and Animal Slaughter," in *Ethics and Animals,* ed. H. B. Miller and W. H. Williams (Clifton, NJ: Humana, 1983), pp. 147–62.

van Hoven, W., "Mortalities in Kudu (*Tragelaphus strepsiceros*) Populations Related to Chemical Defence in Trees," *Journal of African Zoology* 105 (1991): 141–45.

Vertosick, F. T., Jr., *The Genius Within* (Boston: Houghton Mifflin Harcourt, 2002).

Viveiros de Castro, E., "Cosmological Deixis and Amerindian Perspectivism," *Journal of the Royal Anthropological Institute* 8(4) (1998): 469–88.

Volkov, A. G., "Green Plants: Electrochemical Interfaces," *Journal of Electroanalytical Chemistry* 483 (2000): 150–56.

Warren, K. J., "The Power and Promise of Ecological Feminism," *Environmental Ethics* 12(2) (1990): 125–46.

Warren, K. J., *Ecofeminist Philosophy: A Western Perspective on What It Is and Why It Matters* (Lanham, MD: Rowman and Littlefield, 2000).

Warwick, K., *QI: The Quest for Intelligence* (London: Piatkus, 2001).

Weaver, J., *That the People Might Live: Native American Literature and Native American Community* (New York: Oxford University Press, 1997).

Webster, B. D., "Anatomical Aspects of Abscission," *Plant Physiology* 43 (1968): 1512–44.

Weiler, W. E., "Sensory Principles of Higher Plants," *Angewandte Chemie International Edition* 42 (2003): 392–411.

Weisman, A., *Countdown: Our Last, Best Hope for a Future on Earth?* (New York: Little, Brown, 2013).

Wenz, P. S., "An Ecological Argument for Vegetarianism," *Ethics and Animals* 5 (1984): 2–9.

Whitefield, P., *Earthcare Manual* (Hampshire, UK: Permanent, 2004).

Williams, F., *Breasts: A Natural and Unnatural History* (New York: W. W. Norton, 2012).

Willis, R. G., "Introduction," in *Signifying Animals: Human Meaning in the Natural World,* ed. R. G. Willis (London: Unwin Hyman, 1990), pp. 1–22.

Willott, E., "Recent Population Trends," in *Environmental Ethics: What Really Matters, What Really Works,* 2nd ed., ed. D. Schmidt and E. Willott (New York: Oxford University Press, 2012), pp. 526–33.

Winter, J., and M. Titelbaum, *The Global Spread of Fertility Decline* (New Haven, CT: Yale University Press, 2013)

Wu, G. D., et al., "Linking Long-Term Dietary Patterns with Gut Microbial Enterotypes," *Science* 334(6052) (2011): 105–9.

Yanming, Z., Y. Li, L. Jiang, C. Tian, J. Li, and Z. Xiao, "Potential of Perennial Crop on Environmental Sustainability of Agriculture," *Procedia Environmental Sciences* 10 (2011): 1141–47.

Zabel, G., "Peak People: The Interrelationship between Population Growth and Energy Resources," *Energy Bulletin* (April 20, 2009), http://www.resilience.org/stories/2009-04-20/peak-people -interrelationship-between-population-growth-and-energy -resources.

Zerbisias, A., "Cultivating Sustainable Solutions: Weather, Fuel Shortages No Longer an Issue with Vertical Farming," *The Toronto Star* (June 15, 2011), E1.

Zimov, S. A., V. I. Chuprynin, A. P. Oreshko, F. S. Chapin III, J. F. Reynolds, and M. C. Chapin, "Steppe-Tundra Transition: A Herbivore-Driven Biome Shift at the End of the Pleistocene," *The American Naturalist* 146(5) (1995): 765–94.

Zuk, M., *Paleofantasy: What Evolution Really Tells Us about Sex, Diet, and How We Live* (New York: W. W. Norton, 2013).

INDEX

Abram, David
  on ancestor veneration, 78
  on spirits, 52
abscission, 32–33
Adams, Carol
  on factory farming, 5, 43,
    44–45
  on "feeding on grace," 49, 50
  as ontological vegan, 42
  on plants, 44–45
  on the relational hunt, 47–48
  on sacred eating, 48
Adamson, Rebecca, 52
aeroponics. *See* agriculture, vertical
agriculture
  conventional, 34–36, 45, 62, 66–69,
    74, 89, 101, 106, 111, 112,
    122, 128
  ecosystem (*see* perennial
    polyculture)
  organic, 34–36, 62, 92, 101
  vertical, 65–67, 69, 136–37n16
Alvarez, Natasha, 130
ancestor veneration, 77–78, 84,
  87–88
ancient sunlight, 34, 104
animal-industrial complex. *See*
  factory farming
animism, 8, 12, 42, 50–60, 69, 72,
  82–89, 115–17, 121, 126,
  128–29, 133n1, 134–35n5–9,
  137n5, 143–44n18–19
  and anthropomorphism, 55–56

as default conservationist, 51, 57,
  137n5
historical reference to, 51
and kincentrism, 56–57
and the mistaken attribution of a
  human locus, 52
not a matter of faith, 51–52
and other-than-human people, 42,
  54–56, 78, 134–35n7, 135n9
and pantheism, 89
and postmaterialism, 82
and the primacy of the landbase,
  8, 56–57, 61–63, 67, 71, 89,
  101, 122, 124, 136n13
and relationality, 56–58, 135n7
and the sacredness of life, 52
and topocentrism, 56–58
and totemism, 115–17
as "tradition," 51–52
and the transitivity of eating, 87
and utopianism, 56
and who we eat as where we eat,
  63–64, 88
anthropomorphism, 55–56, 127,
  135n11
Appel, Heidi, 18
Aristotle, 12, 17, 96
Australian Aboriginals, 59, 116–17
auxins, 20, 132n7

Bacon, Francis, 3, 90, 131n2
bacteria, 28, 59, 68–69, 75, 83, 88,
  102, 125, 133n11